TURNING RESEARCH AND

turning
RESEARCH
and
DEVELOPMENT
into
PROFITS

a systematic approach

ATTILIO BISIO
LAWRENCE GASTWIRT

amacom a division of American Management Associations

This book was set in Times Roman.
It was designed by Joan Greenfield.
The compositor, printer, and binder was Edwards Brothers.

Library of Congress Cataloging in Publication Data

Bisio, Attilio.
    Turning research and development into profits.

    Includes bibliographical references and index.
    1. Technical innovations. 2. Research, Industrial.
I. Gastwirt, Lawrence, joint author. II. Title.
HD45.B54   658.1'554   78-10239
ISBN 0-8144-5487-9

© 1979 AMACOM
A division of American Management Associations, New York.
All rights reserved. Printed in the United States of America.

First Printing

To the patient ones—
Our wives, children,
and students at
Exxon, Drexel, and Rutgers

preface

The emergence of research and development (R&D) as a major industrial activity has attracted considerable attention and study. Unfortunately, the value of this effort to the R&D manager and the business managers who provide financial support to R&D is questionable. Perhaps the most alarming characteristic of the body of empirical studies is the extreme variance among their findings. Factors judged to be important for R&D in one study are found to be considerably less important, not important at all, or even inversely important, in another.

This variation in results and ideas should not surprise us. R&D is a young industry, grown up from essentially nothing in just 50 years. As it has matured, a series of changes has occurred in its relationship to the supporting and user groups and to government. The relationships existing today are not those of 15 years ago or even, perhaps, of last year.

Industry has been turning more and more to R&D that has direct and immediate benefit to the group(s) providing financial support. The emphasis, and sometimes overemphasis, on achieving an early profit from an R&D activity has led to concern that the future health of the nation's technology and economy is being imperiled by inadequate support of longer-term and riskier R&D.

Meaningful quantitative generalizations on the value of R&D to companies cannot be made. Here, as in other human activities, individuals and organizations are not all equally good performers. However, we can usefully direct our attention to the financial consequences and implications of an R&D program or ideal. That is what this book is about.

Concern over financial consequences rather than technology is, we think, understandable. The costs associated with the translation of an idea into a viable commercial reality are high, and the failure rate is even higher. However, an emphasis on early economic analysis can place those with ideas—be they R&D managers or business managers—at a disadvantage unless they understand the rules of the economic game in their company.

Most of us have spent our working days in a specific functional activity in a company. The narrowness of our assignments, however exciting they are, generally insulates us from the overall economics of our company. Moreover, because our practical knowledge of industry economics as well as our skill in using economic techniques is usually self-learned, often in bits and pieces, we are at a significant disadvantage in selling an idea or even in talking about it to management.

If we are fortunate enough to have a good idea and to work with it, we'll soon get involved in discussions about it with others. As we try to sell our idea, questions will arise. Among those that crop up in a sophisticated corporation are:

What does the venture analysis show?
Have you done any risk analysis?
What's the return for the project on a discounted cash flow (DCF) basis?
Is the project DCF return above the hurdle rate?
At what price will the product sell?

The questions can go on endlessly and often appear overwhelming. They are especially difficult if the party doing the questioning simultaneously supplies answers, particularly since he will rarely if ever communicate his frame of reference. Indeed, in most organizations the relationship among acceptable risk, availability of capital, and the desired rate of return is subjective and elusive, residing in the composite thinking of a small number of decision makers. Their appraisal of an idea is often colored by a significant number of unstated, but important premises against which ideas are tested.

The sellers of an idea need to be able to place both R&D in general and the economic consequences of their specific idea in a perspective that is meaningful for their company and the business managers of that company, and that is relevant to the industry of which their company is a part. This means they must have a good understanding of diverse business planning and financial tools and must also be able to apply them to the analysis of R&D.

The first half of this book serves as an introduction to basic tools and concepts of R&D. Chapters 1–3 deal with the nature of R&D expenditures, R&D successes and failures, and the tradeoffs that must be considered in a business plan. Chapters 4–6 address themselves to basic financial tools that underlie an analysis of the economic consequences of an R&D idea.

Later chapters examine the problems of acquiring and assessing the important data on which major business decisions depend. Chapters 7 and 8 focus on developing an organized discipline for R&D decision making that is similar to that for capital budgeting. Although the major economic decisions associated with new product development are emphasized, the techniques are applicable to other R&D evaluations. Chapters 9–11 consider in depth various approaches to making the "risk and uncertainty" aspects of R&D studies manageable.

This book should be helpful to both the sellers and buyers of R&D ideas. Regardless of their academic and business backgrounds, these parties need a common ground of ideas, concepts, and tools. We hope we have provided one.

Our insights and judgments have been formed by interaction with a multitude of associates and friends. We have had the good fortune to work with some of the world's most challenging, stimulating, and sometimes provoking people. Of the many who come to mind, Bill Catterall, Vern Herbert (retired), Norm Hochgraf, Aimison Jonard (now with the U.S. International Trade Commission), John Pfohl, Bruce Tegge, and Steve Wythe deserve mention. Our thanks also to Howard Oakley for developing the market specification software.

We have particularly benefited from the comments of Jack Crilly, Bob Krebs (retired), Bill Reilly, and Bob Wood in reviewing the manuscript.

We owe a special note of thanks and gratitude to Exxon Corp. and its affiliated companies whose diverse interests, deep involvement in technology, and broad concern with R&D planning have provided us with the basis for much of our understanding. We hasten to add that this book is not an exposition of Exxon policies and operations. We are solely responsibile for the validity of the conclusions and analyses, and also for any errors of fact and deficiencies in logic or in clarity.

Attilio Bisio
Lawrence Gastwirt

contents

chapter 1

The Nature
of
Research and Development

A significant portion of a firm's expenditures is directed at achieving future economic benefits. Many activities not generally considered to be research and development (R&D) are directed at improving a firm's future output and productivity. Among these are the development of new markets, new organizational forms, and new methods. Not surprisingly, over a period of time a firm can change its emphasis to improving existing business areas rather than starting new ones. Or it can stabilize or reduce its expenditures on R&D. An understanding of the nation's pattern of spending on R&D can help us put into perspective the expenditures made by a firm.

The nation's R&D spending was approximately $43 billion in 1977. This represents an increase of 12 percent over the more than $38 billion in 1976. Although the increase is substantial, in real terms it is lower than the 10 to 15 percent per year increases during the period 1953–1966. (R&D expenditures grew from 1.4 to 3 percent of gross national product during the latter period, the increase being fueled by government expenditures.)

Of the $43 billion total, approximately $30 billion was performed by industry, up from $9.5 billion in 1960 and $2.1 billion in 1950. Although industry does about 70 percent of the total R&D work, it provides just under half the funds, with the government providing the rest. Industrial funding of R&D has been climbing in recent years; in 1976 it amounted to about $16 billion (1977 funding by industry is expected to be more than $18 billion, a 16.4 percent jump over 1976 expenditures), up from $5.4 billion in 1968, for a compound rate of growth (in current dollars) of about 9 percent per year.

While national trends are interesting, the critical question is how a company is doing relative to the industry or industries it is participating in. As an example, let us look briefly at the chemicals and allied products industry, one of the major performers of company-funded industrial R&D. This industry, along with electrical equipment and communications, machinery, and motor vehicles, accounted for 75 percent of all industry-supported R&D in the United States in 1977. Various estimates indicate that industry-funded R&D spending for chemicals and allied products amounted to $2.8 billion in 1977. This is about 2.7 percent of industry sales and about 20 percent of before-tax profits.

In the past decade the chemicals industry has mirrored the ebb and flow of R&D funding in U.S. industry in general, although usually with smaller gains. This pattern held true for 1977, when company-funded R&D throughout U.S industry rose 13 percent. The only recent years in which chemicals industry expenditures for R&D grew faster than expenditures for all U.S. industry were 1974 and 1975.

R&D in the chemicals industry therefore took a relaxed upswing in 1977 in comparison with both the overall U.S. R&D expenditures and the chemicals industry's own growth rate in 1974 and 1975. The reason is largely the slowed pace at a few large chemicals companies that experienced vigorous expansion in the past. Among the largest four chemicals companies supporting R&D—Du Pont, Dow Chemical, Union Carbide, and Monsanto—only Monsanto may have had an R&D growth rate higher than the inflation rate for 1977. These four companies account for about 30 percent of all chemicals industry R&D spending. Any pause in R&D by these companies weights heavily on the average.

_____ THE COMPOSITION OF RESEARCH AND DEVELOPMENT

Many activities are encompassed by R&D. It is important at the outset to define these component activities and indicate their relative importance. The definitions employed by the National Science Foundation (whose classifications are the most commonly accepted, although not necessarily the most meaningful) are as follows:

Research and development include basic and applied research in the sciences and in engineering, and the design and development of prototypes and processes. Excluded from this definition are routine product testing, market research, sales promotion, sales service, research in the social sciences or psychology, or other nontechnological activities or technical services.

Basic research includes original investigations for the advancement of scientific knowledge that do not have specific commercial objectives, although such investigations may be in fields of present or potential interest to the reporting company.

Applied research includes investigations directed to the discovery of new scientific knowledge that have specific commercial objectives with respect to products or processes. This definition of applied research differs from the definition of basic research chiefly in terms of the objectives of the reporting company.

Development includes technical activities of a nonroutine nature concerned with translating research findings or other scientific knowledge into products or processes. Development does not include routine technical services to customers or other activities excluded from the above definition of research and development.

The surveys of the National Science Foundation indicate that funds utilized for the development of new or improved products and processes account for by far the largest share of industrial R&D expenditures.

In 1976, for example, U.S. industry performed a total of $26.5 billion in R&D activities. Of this total, slightly more than 4 percent was spent on basic research, just over 19 percent was spent on applied research, and 76 percent was spent on development. This distribution has not changed significantly during the past decade. The figures include industrial R&D that was supported by the government. Considering just the $16.6 billion supported by industry's own funds in that year, about 7 percent went into basic research, 27 percent into applied research, and about 66 percent into development.

All the established surveys of R&D expenditures forecast a continuation of faster growth in R&D spending. The growth in R&D activity has returned to the high rates that characterized the years before 1968, according to the Battelle Memorial Institute's current forecast and analysis of R&D expenditures in the United States. Battelle has estimated a 12.7 percent increase in total industrial R&D performance (private expenditures plus government contracts to corporations) for 1977, or a total of $30 billion, compared with $26.5 billion in 1976.

Business Week's survey of R&D expenditures reveals a wide variation in growth patterns among industry groups and among individual companies within industry categories. According to studies of the inflationary factor in R&D costs by Battelle, for example, companies would have had to increase their R&D budgets by approximately 7.4 percent in 1976 just to keep their R&D efforts up to strength. In fact, most did: all but 6 out of 27 industry categories kept their R&D spending ahead of the inflation rate. In some cases the increases appear to be catchup reactions after years of little change; hence they actually represent a decrease in R&D effort because of the effect of inflation. Makers of leisure time products, from a small base, increased their R&D by a startling 43 percent on average. Steelmakers, traditionally low R&D spenders, reported a big 20 percent increase.

On the other hand, makers of building materials and appliances, reflecting the slow pace of housing starts and sales that prevailed until recently, let their investment in R&D fall behind the inflation rate. Building-materials companies increased their R&D expense only 3.4 percent, while appliance makers registered only a 5.8 percent gain. And one industry, tires and rubber, showed an absolute decrease in R&D spending. Thus, despite the longer-term nature of R&D expenditures, R&D is a discretionary expenditure; and it should not be surprising that spending levels are often responsive to short-term financial forecasts and concerns.

Over the past 20 years many efforts have been made to monitor the total R&D effort in the U.S. economy. In 1951 Congress authorized the National Science Foundation to gather data to track and tally how much companies and their industries were investing in the search for new knowledge and in the development of new processes and products. On a parallel course, and again reflecting the sense of the increasing importance of R&D to management and investors, the accounting profession in the mid-1960s began tightening up its recommended accounting procedures for R&D expenses. Beginning in 1970 the Securities and Exchange Commission required companies to list their R&D expenditures as a separate line item in their "official" annual reports—the Form 10-K statements that all public corporations must file annually.

But the flood of 10-Ks and the annual summaries from the National Science Foundation produced at best a confused flow. On the company level, the data used for the NSF studies are protected by law from disclosure. And although SEC filings are public data, companies did not necessarily file the same data in 10-Ks that they filed in the NSF survey. Since definitions of R&D varied widely among companies, the usefulness of the data was limited. A survey by the Accounting Principles Board concluded that company-to-company differences in the definition of R&D made comparisons meaningless and that year-to-year changes in R&D reporting practices made even comparisons of R&D spending in the same company difficult in different periods. Comparisons of R&D spending across industry lines were almost meaningless. Procter & Gamble Co., for example, properly listed its costly surveys of consumer preferences as R&D expenses, while American Telephone & Telegraph just as properly reported as an R&D cost only its most advanced scientific and exploratory development at Bell Telephone Laboratories, Inc.

The lack of clear definitions for standards and of good guidelines for capitalizing or expensing R&D costs led the accountants to further action. In 1973 the Accounting Principles Board completed a study of the problem under the direction of Oscar S. Gellein and Maurice S. Newman, partners in Haskins & Sells, Inc. In mid-1974 the Financial Accounting Standards Board (which replaced the APB) moved quickly to adopt Accounting Standard No. 2 for R&D.

The financial community reacted strongly to the requirement that all R&D costs be expensed instead of capitalized. Expensing of R&D costs as incurred is a convenient and conservative practice and reduces a potential opportunity for manipulation of short-term earnings. However, expensing R&D costs for purposes of internal company reviews of performance and business planning has some substantial deficiencies. The crux of the problem is that expensing is inconsistent with the long-term-asset nature of the knowledge (or technology) that is purchased through R&D. Typically, only an insignificant fraction of the R&D costs incurred in a given year has any payout that year. The major portion of the cost is incurred with the expectation that the benefits will be seen in future time periods. Expensing of R&D costs, therefore, is contrary to accounting conventions, which attempt to relate costs to the specific revenues generated and which capitalize costs that generate future revenues.

The most important part of Accounting Standard No. 2 (at least from a short-term point of view) is the tighter definition of what can and cannot be listed as R&D expenses. As a result of the new standard, data from different companies are much more closely comparable than they were several years ago. Procter & Gamble's figures on R&D no longer include market research costs; AT&T's data now include several hundred million dollars more in product and process development costs than they did before.

The new accounting standard's definition of R&D follows closely the NSF definitions used by the U.S. Census Bureau, which collects data for NSF. Companies now must report as R&D expenses all costs associated with the search for and discovery of new knowledge that may be useful in developing new products, services, processes, or techniques, or that may significantly improve existing products or processes. They must also report all development costs of significant new products and processes. These costs include the design, construction, and testing of prototypes and the operation of pilot production facilities.

The standard is specific about costs that cannot be listed as R&D. These include routine product improvements, regular or seasonal style changes, testing of production products, and quality control costs. Even engineering followthrough in early phases of commercial production cannot be charged to R&D. Market research and market-testing costs are also excluded, as are legal costs of patent applications and those related to the sale and licensing of patents.

The major difference between the data collected by the Census Bureau and those reported to the SEC on 10-Ks is that the latter show no breakdown between research costs and development costs. There are good reasons for lumping the two together. In its 1965 survey of R&D accounting practices in 245 companies, the American Institute of Certified Public Accountants discovered that there was so much disagreement and legitimate confusion over distinctions between research and development that separating them for accounting purposes usually led to nothing better than guesswork. Simply put, the accountants found that one company's research is another's development.

Virtually all larger companies seem to be trying to follow the new accounting standard as closely as possible. But there is still some confusion, particularly in companies where R&D is a relatively insignificant item. Data from 1974 and earlier may be a bit shaky because companies had to go back and refigure their accounting records on a different basis. But from 1975 on, the numbers should be much more reliable and more useful to analysts, particularly as consistent time series develop over the years.

Surveys from 10-K data conducted by *Business Week* indicate that companies are now filing much the same data to both the Census Bureau and the SEC. The Census Bureau requires reporting from 8,000 companies, covering all companies with 1,000 or more employees and a large statistical sample from smaller ones. The SEC does not require regulated utilities, transportation companies, or companies that spend an insignificant amount—less than 1 percent of revenues—on R&D to report those expenses. Utilities traditionally have not been heavy R&D spenders, but that pattern is changing because of the energy crisis. Some large companies with "insignificant" R&D expenditures that choose not to reveal their precise expenses may actually have sizable R&D operations. For example, R. J. Reynolds, Gulf + Western, and Getty Oil, with combined sales of more than $10 billion, might among them "hide" R&D expenditures totaling $100 million if each spent the maximum of 1 percent.

Ten other companies with sales in excess of $1 billion a year each—including Fluor, Archer-Daniels-Midland, U.S. Industries, and International Minerals & Chemical—have R&D expenses below 1 percent of revenues and elect not to report them to the public. Corporate reasons for not filing these expenditures probably range from a desire to protect information considered proprietary to a desire to avoid the costs of collecting detailed data.

PATTERN OF INNOVATION

The term "innovation," as we will use it, refers to technology actually brought by a company into first commercial use or application, as distinct from simply an invention or a technical prototype. For our purposes, innovation can be viewed as having the following three phases:

Idea generation
Problem solving or development
Implementation and diffusion

The process of innovation is overlaid on a project planning matrix in Figure 1-1. The critical distinction between the phases of innovation and the stop-and-go, sequential nature of business decision making and commercial implementation should be clear.

The process of innovation emphasizes the continuous exchange of information that occurs as a project unfolds and the degree of overlapping (or separation) that is possible for certain project phases. Idea generation involves *synthesis* of diverse information (usually existing, as opposed to original data) about a market or other need and the possible technology to meet that need. Problem solving or development includes setting operational goals and designing alterna-

Figure 1-1.   Project planning matrix.

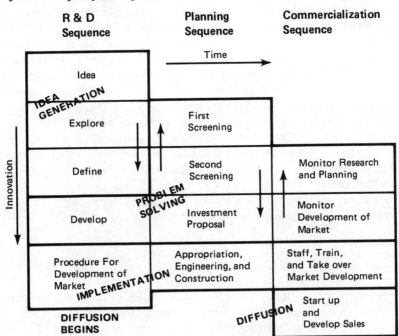

tive solutions to meet them. Implementation consists of the manufacturing, engineering, tooling, and plant and market startup required to bring an original solution or invention to its first use or market introduction. Diffusion takes place in the environment outside the firm and begins after the idea is introduced to the market in some form. It is obvious that many degrees of overlapping are possible.

Clearly, under the NSF definitions R&D is involved in innovation. But the definitions do not help us understand the technological activities of a firm, since they have no essential relation to business objectives. Indeed, R&D in support of any business purpose may include all the NSF classifications of research. Put another way, many different kinds of scientific work can be included in an R&D program, but they should be recognizably and specifically related to a business purpose.

The Industrial Research Institute (IRI) has recommended the following definitions, which rely on the business reasons for performing R&D:

Exploratory research
For advancing knowledge of phenomena of general company interest.
For finding major new high-risk business developments.

High-risk business development
Specific major programs that are aimed at new businesses or other developments of potentially high impact, and that involve higher than normal risk.

Support of existing business
For maintaining or improving the profitability of established businesses.

The current distribution of R&D funding among these three categories is roughly 10 percent, 25 percent, and 65 percent respectively. Perhaps, not too surprisingly, this distribution of funding is similar to the one that obtains for the NSF classifications. IRI has retained the idea of a spectrum of research activities but uses a different criterion from NSF for breaking up the spectrum.

In examining the pattern of innovation in industry, we must start with the idea that a firm or one of its divisions (or product groupings) will show a consistent pattern of innovation over long periods of time.

Consider, for example, the important changes that have occurred in established high-volume product lines, such as incandescent light bulbs, rolled steel, and auto engines. The markets for such goods are well defined, the product characteristics are well understood and often standardized, and competition is based primarily on price. Per-unit profit margins are typically low. The production technology is efficient, equipment-intensive, and specialized to a particular product.

Here the product is defined by the process rather than the converse. The nature of technological change is greatly influenced by the characteristics of the process technology. Change is costly, because in such an integrated system an alteration in any one attribute has many ramifications. In this environment, innovation is typically incremental in nature, with a cumulative effect on productivity. Incremental innovations, such as the use of larger railroad cars and unit trains, have drastically reduced the costs of moving large quantities of materials by rail. While cost reduction is the primary focus, major advances in product performance also result from the sum of numerous small engineering and production innovations.

In such a situation, economies of scale in production and the development of large markets are extremely important. Production facilities are owned by large firms and are located for optimum utilization of materials, labor, transportation, and other factors of production. Even though production costs may be low, the facilities are not flexible in accommodating product change. The system is vulnerable to change demand and technical obsolescence, and the need to maintain production volume to cover fixed costs.

Studies suggest that although minor product variations may easily originate within the large, highly structured firm, new products that require major reorientation tend to originate outside the firm; or, if originated within the company, they may be rejected by it. This is because radical product change involves identification of an emerging or latent need or perception of a new way to meet an existing need; it is an entrepreneurial act.

The stimulus for innovation therefore varies from largely market or needs inputs to largely technology or means inputs as the firm evolves. Initially, when market needs are poorly defined and stated broadly, there is uncertainty about the relevance of outcomes that might be achieved even if R&D resources were committed. The expected value of the R&D is reduced by this target uncertainty, as well as by technical uncertainty. The decision maker has little incentive to invest in risky R&D efforts as long as target uncertainty is high, except as part of a long-term, continuing program. Obviously, such a program could not involve a substantial fraction of the decision maker's resources.

As uncertainty about the appropriate targets for R&D is reduced, R&D projects can become increasingly attractive, and larger and larger R&D investments can be justified. At some point before the cost of implementing technological innovation becomes prohibitively high, and before increasing cost competition erodes margins to levels that cannot support significant R&D expenditures, the benefits of large R&D efforts should reach a maximum. A review of the practice of corporations with high R&D rates provides some support for these observations. These corporations tend to support main business lines that fall neither near the commodity nor the new product boundary, but rather work in a technologically active middle range.

The changing characteristics of a firm's product and process innovation, diversity of product line, production process, organization structure and control, and capacity are brought together in Table 1-1. (The table has been adapted from studies by Thomas Allen, Nicholas Ashford, and Herbert Holloman at the Center for Policy Alternatives, Massachusetts Institute of Technology.) The relationships among the variables are consistent with the findings of many previous studies of innovation and are helpful in explaining many of the dilemmas raised in earlier work.

All the evidence considered shows that a firm typically evolves toward a more rigid process, more homogeneous products, and increased substitution of equipment for labor. This does not mean that evolution continues indefinitely or that it is always in one direction. In a firm that has undergone a substantial degree of evolution, a radical departure from the regular pattern may lead to a reversal of the company's position. That reversal can occur is indicated by studies of work flow structure and organization in firms both before and after major changes. Significant reversals have occurred, for example, in the design and production of new commercial aircraft and in the early major automobile model changes.

In the extreme, continued transition may first be slowed and then halted, resulting possibly in the economic death of a process segment. This can take the form of a geographic migration of production to areas where factor prices of production are low enough to support continued economic vitality. Or it can take the form of absorption and restructuring of product and process, as was the case at the turn of the century with the demise of the gaslight companies.

Although the concept of evolution fits well in most situations, there are exceptions. In some industries evolution has not taken place because the production process initially began at a high state of

TABLE 1-1 Summary of possible relationships between innovation and the evolving structure of the company

	Innovation	Product Line	Production Process	Organizational Control	Kind of Capacity
	Frequent and novel product innovation stimulated by market	High product diversity produced to customer order	Flexible, but inefficient; uses general-purpose equipment and skilled labor	Loosely organized; entrepreneurially based	Small-scale, located near technology source or user; low level of backward vertical integration
	Cumulative product innovations incorporated in periodic changes to model lines Increase in process innovation—internally generated Technology-stimulated innovation	At least one model produced in substantial volumes Dominant design achieved Highly standardized product with few major options	Increasingly rationalized process configuration with line flow orientation, relying on short duration tasks and operative skills of the workforce "Islands" of specialized and automated equipment in some parts	Control achieved through creation of vertical information systems, lateral relations, liaison, and project groups Control achieved through goal setting, hierarchy, and rules as the frequency of change decreases	Centralized general-purpose capacity where scale increases are achieved by breaking bottlenecks Facilities located to achieve low factor input costs, to minimize disruption, and to facilitate distribution
Normal Direction of Evolution of the Firm	Cost-stimulated incremental innovation Novel changes, involving simultaneous product and process adaptations	Commodity-like product specified by technical parameters	Integrated production process designed as a "system" Labor tasks predominantly systems monitoring	Bureaucratic, vertically integrated, and hierarchically organized with functional emphasis	Large-scale facilities specialized to particular technologies; capacity increases achieved only by designing new facilities

development. This appears to be the case in many chemical or other continuous-flow processes where existing standard-process equipment can be used to produce a new product virtually at will. The exception extends beyond the pure continuous-process industries to products like cigarettes and foodstuffs where the available production technology defines the mode of operation.

A more interesting exception to the notion of evolution is found in areas where development has simply failed to come about—as in home construction, nuclear plant construction, and selected aspects of health care delivery. In each of these areas experimental programs to stimulate cost reductions, greater standardization, or other transitions have been undertaken through government and private sponsorship with little success. Firms that have low capital intensity or low R&D inputs and that are in an early period of development tend to see factors that impede product standardization and market aggregation as the most important barriers to evolution. Larger firms tend to rank disruptive factors like uncertainty over government regulation and protection of existing investments as more important.

chapter 2

The Risks and
Benefits
of
Research and Development

Along with the rapid growth in R&D spending, and perhaps in part as a consequence of it, widespread concern is developing over R&D. However, the nature of the concern varies with the point of view of the observer.

Despite considerable confusion and disagreement over the problems of corporate R&D, Arthur D. Little, Inc. reports virtual unanimity among industrialists on one point: the area is one of increasing concern. This concern is evident in a variety of symptoms: worry over the growing costs of R&D; the feeling that corporate R&D has not "produced" new products and processes at a rate commensurate with expenditures; questions about the organization and composition of R&D activities; and questions about methods for evaluating the accomplishments of R&D. Underlying these symptoms is a disillusionment with the attitude that prevailed from the late 1940s through the early 1960s, when it was widely held that "good research" in the corporate setting would of itself meet industry's needs for innovation.

Industry's concern is evident in the increasing number of workshops, symposiums, and conferences on research management offered by such organizations as Arthur D. Little, Inc., the American Management Associations, and Battelle Memorial Institute. The comments that follow are typical of industry's attitudes in recent years:

1965

. . . Many leaders of industry are growing uneasy about the soaring bills they are paying for research and development. The total, now $6 billion a year, is double what it was a decade ago. Top managements—often prodded by stockholders and financial houses—are increasingly asking: Are we running R&D as efficiently as we could? Are we really getting our money's worth?

An emphatic "no" comes from Monsanto, the nation's third-largest chemical company. . . . Board Chairman Charles Allen Thomas recently declared: "I feel we are in danger of being run down by our own research and development programs. . . . We have reached a point of indigestion . . . and I am of the opinion that the productivity of this effort has fallen off, whether you measure it on a man-hour basis or on a dollar basis." The nation's R&D, he concluded, "is now stumbling in a plethora of projects, sinking in a sea of money, and is being built on a quicksand of changing objectives."*

1977

. . . In fact, the changes at Du Pont are interesting mainly because of the company's size, its technical depth, and its historical financial health. If these strengths once seemed to insulate Du Pont R&D from market hazards, it is clear that they no longer do. Now, says [Chairman Irving S.] Shapiro, Du Pont's R&D effort "must reflect the fact that we are scrambling for capital and scuffling for profits."

In recent years, Du Pont's R&D budget has declined in relation to the size of its business. The company's R&D outlay this year is budgeted at 4.2% of sales. That percentage is higher than for most other big chemical companies: Celanese, Dow, Monsanto, and Union Carbide all put less than 3.5% of sales into R&D in 1975. A decade ago, however, Du Pont put 8% of its $3.5 billion in sales into R&D, and as late as 1971, when sales were $4.4 billion, the figure was 5.7%. "The fundamental change," says Shapiro, "is that we are now holding research expenditures at a relatively fixed level."

The company's aim is to adjust that figure upward to match inflation, and thus keep some 3,900 scientists and engineers working in 125 separate

*Hubert Kay, "Harnessing the R&D Monster," *Fortune* (January 1965), p. 160.

Du Pont laboratories. But as sales rise, says Senior Vice President Richard E. Heckert, "research will continue to shrink as a percentage of sales."*

A decade may pass, but the concerns remain.

In reaction, there is a concern that management is becoming supercautious and unwilling to gamble on anything short of a sure thing. The impact of such thinking obviously varies from company to company. Among companies that are rooted in innovation—instrumentation, computers, and electronics, for instance—important products and processes continue to churn out of the research lab. In other industries—steel, chemicals, paper, packaged goods, and automobiles—product innovation has leveled off or dropped. Yet nearly all leading companies, whether strong innovators or not, share one attribute: their corporate approach to innovation is becoming highly structured and more tightly controlled. Among the key shifts in direction are these:

o Performance demands on new products are getting stiffer. Compared with a few years ago, products stay longer in research and testing. Fewer dribble out to the market, and those that do must often pay back their investment much earlier than before, sometimes in only one-third to one-half the time. This is because market pressures have cut many product life cycles by 40 percent to 60 percent over the past ten years.

o Top management is getting deeper into key decisions on innovation and new product investment.

o Some of the innovative thinking that went into new products is being redirected into existing products and processes. While that may contribute to lag in technological innovation, it can stimulate more marketing innovation. The redirection of thinking shows up in better quality, packaging, and "positioning" of existing products. It also shows up in important process innovations and new manufacturing efficiencies that help increase productivity of existing systems and hold overall costs and prices down.

For an economy built on innovation, the implications of a technological slowdown or redirection could be significant.

_____ THE BEST ODDS

Many executives examine, with some alarm, what might be called the R&D utilization record—the percentage of new product develop-

Chemical & Engineering News (January 17, 1977), pp. 14–15.

ment projects that ultimately result in commercial products. Available data indicate that this percentage is low.

Surveys by Booz, Allen & Hamilton, Inc. indicate that it takes 40 new product ideas to generate 13 development projects. Of the 13 projects that undergo development, two will result in commercial products and one will be successful. Other recent surveys indicate that among 51 leading firms, it took 58 new product ideas to generate seven development projects and, ultimately, one successful new product. According to the Commercial Development Association, of every 540 ideas proposed for new industrial chemicals, 92 are selected for preliminary laboratory evaluation, eight reach the development phase, and one is commercialized. *Business Week* estimates that for every ten chemicals projects that leave the laboratory and enter development, only one reaches commercial production.

The failure rate for consumer products is just as significant. Here are the figures from different surveys over the period 1960–1975 for new food and drug items:

Nielsen	53% failed
Business Week	50–80% failed
Rosen	Over 80% failed
Dodd	Over 80% failed
Helene Curtis	43% failed
United Kingdom	Over 40% failed

The figures for new packaged consumer goods are:

Angelus	Over 80% failed
Booz, Allen & Hamilton	37% failed
The Conference Board	40% failed
Ross Federal Research Corp.	80% failed

The highest failure rate known to us is 90 percent, reported in a U.S. Department of Commerce study.

The results of such surveys must be considered with caution, since the terminology is extremely loose. What is meant by an "idea," by "preliminary laboratory evaluation," by a "development project"? Indeed, what is meant by a "new product"? These are more or less subjective concepts, and it is dangerous to lump together data from many firms.

Despite the imprecision, we can conclude that the odds that any given research project will result in a successful product are not one out of two or even one out of three (probabilities with which business executives may feel reasonably comfortable); they are more like an order of magnitude poorer. This conclusion is corroborated by spokesmen for leading companies. In the early 1950s Crawford H. Greenewalt, then president of Du Pont, discussed his firm's record as follows: "With all of the many skills available to a large enterprise, with all of the fine equipment, with men of the highest scientific attainment, with able and experienced management, a given research project at Du Pont has about one chance in twenty of becoming a commercial success."*

Today, some 25 years later, many research managers we know would say, "Chances of winning through to a successful new product today are becoming less and less." But what of specific companies known as technology leaders? Are their records better than the average?

An insight into the Du Pont record was given by Chaplin Tyler of that company in 1962. Tyler discussed a memorandum, written nearly seven years earlier, on new product research projects then in progress that looked particularly significant. All together, 20 products were listed in various stages of development, from laboratory exploration to field evaluation. By 1962, six of the products (30 percent) were in commercial-scale production; seven (35 percent) were still in development and appeared to have at least an even chance of success; the remaining seven (35 percent) were "disappointingly slow." The probability of success for products in this third group was seen as "not good."

Tyler's view was that by January 1966 half of the original 20 product candidates would attain commercial status. The anticipated ten commercialized products were the survivors, Tyler estimated, of perhaps 200 original research projects. It should be kept in mind that the 20 projects under discussion were unrepresentative of the total population: they were chosen because they were "particularly significant." Similar data have been cited by Minnesota Mining & Manufacturing. For every 100 laboratory starts, 3 M has an average of 33 technical successes; and only three of these ever become commercial successes.

As a result of this low R&D utilization record, much of the money spent on new product development is considered wasted. For example, Booz, Allen & Hamilton studies indicate that 70 percent of new product expenditures are spent on products that are not successful. In the development stage alone, in which R&D has the primary role, about

Chemical & Engineering News (November 10, 1952), p. 30.

eight out of ten new product dollars are spent on products that will not be justified in terms of their ultimate commercial value. Expressed in a more dramatic way, four out of five scientists and engineers engaged in technical development are working on new products that will not achieve commercial success.

Is the productivity record really poor enough to warrant the concern displayed over it? It is not at all clear that this is the case. Corporate annual reports are replete with statistics like these: "*x* percent of our current sales is contributed by new products developed over the last ten years"—and *x* is often on the order of 50 percent. It is, accordingly, dangerous to equate a *low* R&D utilization record with a *poor* R&D record. Certainly, the utilization statistics are meaningless by themselves. We must also consider the influence on the course of the business of the new products brought forth from R&D and the *overall* profitability of R&D expenditures—that is, the profits from the one success compared with the costs of the 10, 20, or 30 failures incurred along the way.

_____ WHY THE FAILURE RATE?

R&D has undergone 50 years of technological development. It has a growing body of psychological and mathematical hypotheses (if not theory or, in some cases, confirmed facts or laws), a reasonably complete literature, and excellent journals, facilities, and people. Why, then, is there such a high rate of new product failure? The reasons various investigators have cited for failure are given in Table 2-1. To facilitate comparison between the studies, we have taken some liberty with the terminology. For example, in the Hopkins and Bailey study as originally published, the principal causes of failure for new products and services were given as shown in Table 2-2.

Obviously, a combination of factors is responsible for new product failure. High on any list are the factors of poor planning, poor timing, and the tendency to let enthusiasm override caution.

The great majority of studies on innovation have concentrated on a company's success. To our knowledge, only three studies have clearly addressed themselves to distinguishing the characteristics of commercially successful projects from those judged unsuccessful:

James Utterback at the Center for Policy Alternatives, Massachusetts Institute of Technology, studied 164 projects in Europe and Japan.

Christopher Freeman at the University of Sussex considered 58 innovations.

Albert Rubenstein at Northwestern University studied 144 cases from eight firms in the United States.

TABLE 2-1   Surveys of new product failure

Reason for Failure	A. R. Abrams, Inc.	Angelus	Booz, Allen & Hamilton, Inc.	Constandse	Diehl	Hopkins and Bailey	MacDonald	Miles	TOTAL
Lacked meaningful product uniqueness*	X	X	X	X	X	X	X	X	8
Poor overall planning**	X	X		X	X	X	X		6
Wrong timing	X	X	X	X		X			5
Enthusiasm crowded out facts				X	X	X	X	X	5
Product defective	X	X				X			3
Product lacked a champion in company					X				1
Company politics					X				1
Unexpected high product cost						X			1

*If there was a difference, its value to potential buyers was overestimated.
**Includes such factors as poor positioning, poor segmentation, underbudgeting, poor overall themes, and overpricing.

TABLE 2-2   Hopkins and Bailey study of causes of new product failure

Cause of Failure	Percentage of Companies Citing
Inadequate market analysis	45
Product problems or defects	29
Lack of effective marketing effort	25
Higher costs than anticipated	19
Competitive strength or reaction	17
Poor timing of introduction	14
Technical or production problems	12
All other cases	24

Findings from these studies are compared in Table 2-3. The most striking characteristic of successful projects is that they had little or no initial difficulty in marketing; in contrast, nine out of ten failures experienced difficulty in initial market attempts. This finding is strongly supported by the MIT and Sussex studies. Successful innovations had fewer aftersales problems, required less adaptation by users, had fewer technical "bugs" or unexpected adjustments in production, and needed fewer modifications resulting from user experience after commerical sales.

A host of other competitive and market-related factors also seem

TABLE 2-3   Factors related to project success

Description of Factor	MIT	Northwestern	Sussex
Market-Oriented Factors			
No initial difficulty in marketing the product.	***		***
Project perceived to have great competitive advantage.	**	**	
Specific competitive stimulus for project.	*		*
Need recognized among users.			
Project intended for a specific user or end product.	**		**
Project members had frequent contact with users.			**
Resource Factors			
Project considered fairly or highly urgent.	***	***	
Project initiated by top management of firm.		*	**
Organization Factors			
Structured planning process.		***	
Government Factors			
Regulatory constraints perceived to be highly significant.	**	**	

***A very strong relationship.
**A strong relationship.
*A weak relationship.

to be significant. Successful projects had a great advantage over competing approaches or products in a key feature, or moderate advantage in several features important to the user, and were frequently directed to areas where the firm had specific competitors. For successful projects a need was recognized to exist among users and was identified before a solution (or technical opportunity) was available. In addition, successful projects were intended more often for a particular user or user specification, and project members maintained more frequent contact with users. Another striking finding is that successful projects seemed to be related to a fairly or highly urgent problem faced by the firm and were more frequently initiated by the firm's top management (president or a management committee).

Both the Northwestern study of U.S. firms and the MIT study of firms in Europe and Japan discovered a negative relationship between adequacy of patent protection and commercial success. Unsuccessful cases in which respondents considered that an easily defensible patent could be obtained outnumbered patentable successes nearly three to two. This amply illustrates the fact that a relationship between two such variables does not imply a causal connection. It may be that easily patented innovations are technology- rather than market-stimulated, complex and/or narrowly focused, and therefore less likely to be successful. Successful projects may be incremental product variations with little market or technical uncertainty. There is some evidence in the Sussex study to support this view. Conversely, it may be that successful projects face greater competition and many competitive alternatives and are thus more difficult to patent. There is also some evidence to support this view: in all three studies competitors' patents were reported to be a significant problem in more of the success cases.

RETURNS FROM INDUSTRIAL R&D

The enormous difficulties, both conceptual and practical, in estimating the return from industrial R&D are all too obvious. However, the few estimates that are available suggest that the returns are substantial. (All returns described below are expressed on a before-tax basis.)

Edwin Mansfield and his associates at the Wharton School recently studied the returns from 17 innovations in a variety of industries and firms of quite different sizes. Most of the innovations were judged to be of average or routine importance. The median rate of return was 25 percent. Mansfield also obtained detailed data for the period 1960–1972 from one of the nation's large firms. Each year the firm made a careful inventory of the benefits from R&D on its income stream. The average rate of return from this firm's total investment in innovative activities was 19 percent.

The foregoing results pertain to the average rate of return. That is, they are the average rates of return from all amounts spent on the relevant R&D. For many purposes, a more interesting measure is the marginal rate of return, or the rate of return from each additional dollar spent. In investigations based on econometric estimates of production functions, Mansfield and Jora Minasian independently estimated the marginal rate of return from R&D in the chemicals and petroleum industries. Mansfield's results indicated that the marginal rate of return in the petroleum industry was 40 percent or more; in the chemicals industry it was about 30 percent when the technical changes resulted in new production facilities but much less when only changes in techniques were involved. (Economists speak of this distinction as capital embodied versus disembodied.) Minasian's results indicated about a 50 percent rate of return on marginal investment in R&D in the chemicals industry.

High average rates of return are not limited to process industries. Mansfield's studies have shown substantial returns in other industries as well. For example:

Food, apparel, and furniture	15%
Agriculture	35–170%
Selected other industries	15–55%

Nestor Terleckyj has used econometric techniques to analyze the effects of R&D expenditures on productivity in 83 manufacturing and nonmanufacturing industries during 1948–1966. In manufacturing, the results indicate about a 30 percent rate of return from an industry's R&D based only on the effects on its own productivity.

Zri Griliches carried out an econometric study, based on data for almost 900 firms, to estimate the rate of return from R&D in manufacturing. He found the private rate of return to be about 17 percent. The figure was much higher in chemicals and petroleum and much lower in aircraft and electrical equipment. Interestingly, Griliches found that the returns from R&D seemed to be lower in industries where much R&D was federally financed. This could be a result of either the longer-range nature of the project or perhaps poor commercial prospects. In either case, a company might seek federal funding if it were available.

In a study covering the economy as a whole, Edward Denison concluded that the rate of return from R&D was about the same as the rate of return from investment in manufacturing. His estimate of the returns from R&D was lower than the estimates of other investigators, perhaps because of his assumptions regarding the time period required to realize the benefits. In his 1970 presidential address to the American Economic Association, William Fellner estimated the average rate of return from technological progress activities to be "substantially in excess" of 13 or 18 percent, depending on the cost base, and "much higher than the marginal rate of return from physical investments."*

There is one important limitation to these studies, despite our positive overall assessment. Econometric and statistical studies showing a significant positive relationship between R&D and sales growth, profits, and productivity cannot be used as evidence of cause and effect. All may be dependent variables with a third factor the real independent or causal variable. For example, if a firm shows a positive correlation between R&D and sales, both of these may be the result of an enlightened (new) and progressive management. The multiple influences on a firm's sales growth and profits (let alone the economy's growth and productivity) and the complex interactions among them make it difficult to distinguish the particular contributions of one factor such as R&D.

Despite this frailty, it is remarkable that so many independent studies based on so many types of data result in so consistent a set of conclusions. To sum up, practically all the studies carried out to date indicate the *average* rate of return from R&D to be high and at least equal to that from investment in capital goods.

American Economic Review (March 1970).

chapter 3

Strategic Planning and Research and Development

The past decade has seen growing recognition that research and development in diversified companies of all sizes involves tradeoffs among competing opportunities and strategies. During this period the combination of more complex markets, shorter product and process life cycles, and social, legal, and government trends have increased the premium on minimizing the degree of risk in the overall mix. More recently, managers have had to cope with severe resource constraints, stemming partly from weaknesses in the capital markets and a general cash shortage and partly from the traumas of the energy crisis, environmental constraints, potential product liability, and inflation.

Among the manifestations of the new climate for R&D are skepticism toward the value of full product lines, unwillingness to accept the risks of completely new products and processes, emphasis on profit growth rather than volume growth, and active product elimination programs. Yet managements cannot afford to turn their backs on opportunities for change and attempt to survive simply by doing a better job with established products and services. Eventually all product categories become saturated or threatened by substitutes, and diversification becomes essential to survival.

Consumer goods companies especially are feeling this pressure as the productivity of line extensions or product adaptations directed at narrow market segments declines. Also, the possibility of regulatory actions directed at products, such as restrictions on aerosols and cyclamates, points up the risks of having a closely grouped product line. More than ever, long-run corporate health is going to depend on the ability of management to juggle the conflicting pressures of diversification and consolidation.

If a company obtains promising results from a piece of research, it must decide whether to develop the idea to full-scale commercialization. Alternatively, if a firm has ideas for several projects but the financial and technical resources to develop only a few of them, it must determine how the projects are to be selected. Other decisions also occur throughout the development stages of a project: whether to carry on or stop and cut possible losses in the light of changing circumstances; and, if the decision is made to carry on, what to do next.

These decisions are facilitated if a project can be measured and evaluated. An evaluation consists of weighing the effort to develop an idea to the commercial level against the eventual benefits expected from it. Experienced managements may make such an evaluation qualitatively or even intuitively; but the more complex a project is, the more difficult and less effective subjective evaluation becomes. More systematic, quantitative methods are required.

To compare development effort with ultimate benefits quantitatively, both must be measured in the same units, and the universal units are money. Thus price tags are attached to all phases of the project. The economic evaluation is based on estimates—estimates of the cost of bringing the project to the point of profitable exploitation and of the profits that will then be obtained.

In an economic evaluation of a project in its early stages, it is unrealistic to consider only the cost of research and development without taking into account the other activities needed for successful exploitation. Perfecting a piece of technology may be of little value unless financial and other resources are available at the right time to exploit it so that the financial benefits can be attained. R&D must be considered only in terms of the benefits it makes possible for the entire organization. In these terms, an R&D project can be regarded as the first link in a chain that extends through to commercialization, production, and marketing.

All the expenditures necessary are therefore included in the evaluation: the cost of the applied research and development program itself and the cost of capital investment in plant, machinery, and other facilities as well as any related expenses. Also required, in order to forecast profitability, are data on operating costs and product price and demand. The success of the evaluation depends on the cooperation of many groups, including research and development, marketing and market research, design engineering, construction, and production. This is not a simple task, but it can be done, provided those making the effort understand how R&D affects a company's strategies.

STRATEGIC PLANNING AND R&D PLANNING

There are as many concepts of strategy as writers on the subject. Several of the more useful definitions of strategy for our immediate purposes are:

Decisions today that affect the future (not future decisions).

Major questions of resource allocation that determine a company's long-run results.

The calculated means by which a firm deploys its resources—personnel, machines, and money—to accomplish its purpose under the most advantageous circumstances.

A competitive edge that allows a company to serve customers better than its competitors.

The broad principles by which a company hopes to secure an advantage over competitors, an attractiveness to buyers, and a full exploitation of company resources.

Following these definitions, the goal of strategic planning is a long-run program, as David Dollat has written, "that will produce an attractive growth rate and a high rate of return on investment by achieving a market position so advantageous that competitors can retaliate only over an extended time period at a prohibitive cost."*

Most strategic-planning programs show a distinct family resemblance, although the specifics obviously vary greatly. These specifics usually include statements of the following:

The mission of the program.

The desired future position the program will help the corporation attain, including profitability, sales, market share, and efficiency objectives.

The key environmental assumptions and the opportunities and threats.

The strengths, weaknesses, and problems of the program and the major competitors.

The strategic gap between the desired and forecast position of of the program.

Actions to be taken to close the gap—the strategy.

The required resources and where they can be obtained, including human as well as financial resources such as net cash flow, the equity base, and debt capacity.

These are the main elements of the planning process that are relevant to R&D, leaving aside detailed implementation plans and contingency plans, which state in advance what modifications will be made if key assumptions about the environment or competitors turn out to be false.

The planning process will involve different concerns at various levels within a company. For companies with divisional organizational structures, for example, the concerns of each level will differ in accordance with the decentralization rationale behind the organization structure:

Organizational Level	Primary Strategic Concern
Corporate headquarters	Management of business area involvements by maintaining a desirable "portfolio" of businesses
Division	Strategic management of the business the division is in

*David T. Kollat, Roger D. Blackwell, and James F. Robeson, *Strategic Marketing* (New York: Holt, Rinehart and Winston, 1972), pp. 21–23.

Department   Functional decisions such as mar-
                    keting and production strategy

In line with the different strategic concerns, there will be three levels of planning.

Planning systems don't happen; they evolve. A number of case studies indicate the following significant differences in "new" versus "mature" planning systems with respect to how critical issues are handled:

| | *Formal Planning* | |
Issue	*New Systems (First Year)*	*Mature Systems (Later Years)*
Communication of goals by corporate level to divisional level	Not explicit	Explicit
Initiative for divisional level's goals	Bottom up	Negotiated
Focus of division's planning activity	Financial	Strategic
Linkage of plans to budgets	None	Tight

In mature planning systems, R&D is a critical component in establishing the strategic options open to a company.

A systematic procedure for generating and choosing strategic alternatives leads inevitably to consideration of the "product life cycle" and "product portfolio" concepts. Experience has taught us that a wide variety of actions are appropriate for different companies in different industries at different times. No one approach applies to all situations. In fact, we often wonder whether a single approach is optimum for any two companies.

The greatest weakness of many plans is the lack of alternatives presenting different approaches and outcomes. Too often, the strategy favored by top management ignores the life cycle of the company and its products and the interdependence among products. Companies, as a result, pursue strategies that are simply not appropriate for them. One of the major problems facing many U.S. companies today is that they are operating in a mature or aging sector of the economy. Growth may not be a viable alternative, but short-term profits must be—and that has serious implications for R&D.

_____ PRODUCT LIFE CYCLE

The product life cycle has been viewed as another example of a time-dependent intermediate-term forecasting model, based on a biological analogy. This view is misleading in that the model has been an aid in planning and policy formulation rather than forecasting. In planning studies, the life cycle concept has been championed by Arthur D. Little, Inc.

The characteristic product life cycle curve shown in Figure 3-1 finds strong theoretical support in Everett Rogers' theory of the diffusion and adoption of innovations. Essentially, the concept implies that a new product meets initial resistance in marketing and is purchased by only a limited segment of the buying population. Later, as the product's performance and value become known and communicated, a larger segment of buyers adopts it and sales begin to increase at a faster pace. Eventually, the rate of growth decreases as the proportion of adopters approaches a maximum, with most sales representing repeat purchases. The rate of adoption remains constant throughout the mature phase and diminishes in the aging phase. The link between Rogers' theory and the life cycle concept becomes obvious if we consider that the logistic curve usually employed to represent the life cycle is the cumulative equivalent of the normal density function, which is precisely the shape of Rogers' adoption function.

While the relationship between the life cycle and the theory of

Figure 3-1. Hypothetical life cycle patterns

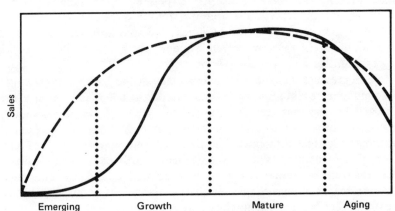

adoption provides a plausible rationale for the life cycle model, the choice of a logistic curve between the introductory and mature periods is an unnecessary restriction. The diffusion of many new products resembles an exponential curve (the dashed line in Figure 3-1), especially if the item is not a dramatic innovation and if its entry into the market is supported by significant promotion efforts.

The interpretation of the life cycle as having four main stages and a specific sales pattern is far from universal. Some researchers have attempted to develop a taxonomy of different life cycles. William E. Cox, in a study of 258 ethical drug brands, identified six patterns and found that for more than 50 percent of these a fourth-degree polynomial best fixed the historical data. His results are similar to those of John Hinkle of A. C. Nielsen Co., who identified a "recycle pattern" for certain product categories. The recycle model in Figure 3-2 illustrates their analyses.

Figure 3-2. Nielson-Cox recycle model

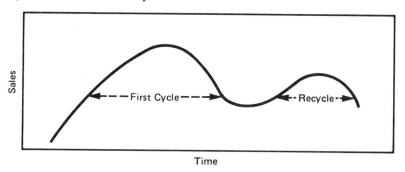

Regardless of the model, not all segments of the cycle take the same amount of time, nor are the life spans the same in all industries. Similarly, companies do not travel right on the curve. A company can regenerate itself or a product (the recycle model), backing up the curve apparently from the mature to the growth phase.

Arthur D. Little, Inc., in studying a large number of companies and constructing strategic profiles of their business, has identified a number of generic strategies involving financial, marketing, manufacturing, R&D, and management actions. For example:

Backward integration	Development of overseas business
Excess capacity	Forward integration
Licensing abroad	Initial market development
Market penetration	New products/same markets
Technological efficiency	Same products/new markets
Traditional cost cutting	New products/new markets

If the different strategies are superimposed on the product life cycle, some turn out to be more appropriate than others at a given time. For example, integrating backward to the manufacture of parts and even the raw materials for a product deserves particular consideration during the growth and mature phases. Building excess capacity for a product must be an integral part of the company's planning for growth. Obviously, some of these strategies will also have serious implications for R&D direction, emphasis, and funding.

_____ PORTFOLIO STRATEGIES

The product portfolio approach to marketing strategy has gained wide acceptance among managers of diversified companies. Just as the life cycle approach is favored by Arthur D. Little, Inc., so the portfolio approach has been widely championed by the Boston Consulting Group. Managers are attracted to the portfolio approach because of its intuitively appealing concept that long-run corporate performance is more than the sum of the contributions of individual profit centers or product strategies. The product portfolio analysis can suggest marketing strategies to achieve a balanced mix of products that will produce the maximum long-run effects from scarce cash and managerial resources. The concept also lends itself to a simple matrix representation that is easy to communicate and comprehend.

Common to all portrayals of the product portfolio is the recognition that the competitive value of market share depends on the structure of competition and the stage of the product life cycle. The relationship can be portrayed by a cash quadrant or share growth matrix, as shown in Figure 3-3. Each product is classified jointly by rate of present or forecast market growth (a proxy for stage in the product life cycle) and a measure of market share dominance.

The arguments for the use of market share are familiar and well documented in the studies of the Market Science Institute. Their project on the ongoing profit impact of market strategies (PIMS) gives support to the proposition that market share is strongly and positively

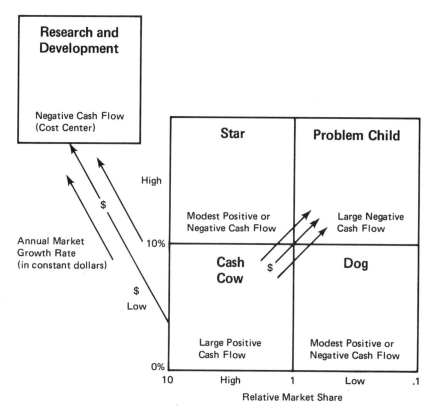

Figure 3-3. The product portfolio in the market share growth matrix

correlated with product profitability. This theme is varied somewhat in the Boston Consulting Group approach by the emphasis on relative share, as measured by the ratio of the company's market share to the share of the largest competitor. This measure is reasonable, since the strategic implications of a 20 percent share are quite different if the largest competitor's is 40 percent or if it is 5 percent. Profitability will also vary, since both the experience-curve concept and that of scale of operation suggest that the largest competitor should be the most profitable at any prevailing price level.

Each of the basic categories in the share/growth matrix in Figure 3-4 implies a different set of strategy alternatives. Obviously, our comments below must be regarded only as generalizations; in application, they must be specific to given products in a given industry at a given time.

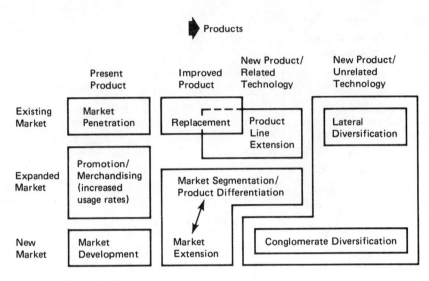

Figure 3-4.   Growth vector alternatives

STARS   Market leaders that are also growing fast will have substantial reported profits but need a lot of cash to finance the rate of growth. The strategies for Stars must be designed to protect the existing market share level by reinvesting earnings in the form of price reductions, product improvement, better market coverage, and production efficiency increases. Particular attention must be given to obtaining a large share of the new users or new applications that are the source of growth in the market. Management could elect, instead, to maximize short-run profits and cash flow at the expense of long-run market share. Such an approach is risky because it is predicated on a continuing stream of product innovations and may deprive the company of a Cash Cow that is needed in the future.

CASH COWS   The combination of slow market growth and a substantial position in the marketplace usually results in substantial net cash flows. The amount of cash generated by Cash Cows often exceeds the amount required to maintain market share. Management should consider strategies directed toward maintaining its position in the market, including investments in technological leadership. Pressure to overinvest through product proliferation and market expansion must be resisted unless prospects for expanding primary demand are unusually attractive. Instead, management should consider using excess cash to support research activities and growth areas elsewhere in the company.

DOGS   Since there usually can be only one market leader and since many markets are mature, the greatest number of products fall into the Dogs category. Such products are usually at a cost disadvantage and have few opportunities for growth at a reasonable cost. Their markets are not growing, so there is little new business to compete for, and market share gains will be resisted strenuously by competitors.

The product remains in the portfolio because it shows (or promises) a modest book profit. This accounting result can be misleading, because a substantial part of the cash flow may have to be reinvested to maintain the company's competitive position and to finance inflation. Unfortunately, in these situations individual investment projects (especially those designed to reduce production costs) generally show a high return. However, the competitive situation may be such that this return cannot be realized. In addition, management should consider the potential hidden costs of unproductive demands on management time (and consequent missed opportunities) and low personnel morale resulting from lack of achievement.

The need to take positive action with a Dog can become urgent. The search for alternatives must begin with attempts to alleviate the problem without divesting. All cost-cutting activities need to be examined, including a conscious cutback of R&D and all other support costs so as to maximize the product's profitability over a short lifetime. Alternatively, the product or business can be reparceled. If these possibilities are unproductive, the options are divestment (after making the product as attractive as possible), liquidation, and finally, if need be, abandonment.

PROBLEM CHILDREN   The combination of rapid growth and poor profit margins creates an enormous demand for cash. If the cash is not forthcoming, as growth slows the Problem Child will become a Dog. The basic options are fairly clear-cut: either invest heavily to get a disproportionate share of new sales or buy an existing share by acquiring competitors and thus move the product toward the Star category. If these options fail, get out of the business using some of the Dog methods discussed previously.

The long-run health of the corporation depends on having some products that generate cash (and provide acceptable reported profits) and others that use cash to support growth. Among the indicators of overall health are the size and vulnerability of the Cash Cows (and the prospects for the Stars, if any) and the number of Problem Children and Dogs. Particular attention must be paid to those products with large cash appetites. Unless the company has abundant cash flows, it cannot afford to sponsor many Dogs or Problem Children at one time. If resources (including debt capacity) are spread too thin, the company will wind up with too many marginal products and will weaken its ability to finance promising new product entries or acquisitions in the future.

Yet corporate objectives have many more dimensions that require consideration. This point was recognized by Seymour Tilles in 1963, in one of the earliest discussions of the portfolio approach. It is worth repeating as a caution against a myopic focus on cash flow considerations. Tilles' point is that an investor pursues a balanced combination of risk, income, and growth when acquiring a portfolio of securities. He further argues that "the same basic concepts apply equally well to product planning."* The problem with concentrating on cash flow to maximize income and growth is that strategies to balance risks may not be explicitly considered.

A company must avoid excessive vulnerability to a specific threat from one of the following areas:

The economy (for example, business downturns)
Social, political, and environmental pressures
Supply continuity
Technological change
Unions and related human factors

It follows that a firm should direct its new product search activities at several different areas to avoid intensifying its vulnerability. Thus many companies in the power equipment business, such as Brown Boveri, are in a quandary over whether to meet the enormous resource demands of the nuclear power equipment market, which is much more vulnerable to economic, technological, and other changes than is a market such as household appliances.

The desire to reduce vulnerability can be a reason for keeping, or even acquiring, a Dog. Firms may integrate backward to insure a supply of highly leveraged materials. Similar arguments can be made

Harvard Business Review (July–August 1963).

for products acquired for "intelligence reasons." For example, a large Italian knitwear manufacturer acquired a high-fashion dress company that sells only to boutiques to help follow and interpret fashion trends. Similarly, because of the complex nature of the distribution of lumber products, some suppliers have acquired lumber retailers to learn about patterns of demand and changing end-user requirements. In both these cases, the products and businesses were acquired for reasons outside the logic of the product portfolio and should properly be excluded from such an analysis.

DESIRABLE ALTERNATIVES

At the broad level of R&D strategy, the basic issues are the growth vector, or the direction the firm is moving within the chosen product market scope, and the emphasis on innovation versus imitation.

There are almost an infinite number of possibilities for growth vectors. The basic alternatives are summarized in Figure 3-4. The strategies depicted are not mutually exclusive; indeed, various combinations can be pursued simultaneously in order to close the strategic gaps identified in the overall planning process. Furthermore, most of the strategies can be pursued either by internal development or by acquisition, coupled with vertical diversification (either upward toward a business that is a customer or downward toward a business that is a supplier).

The choice of growth vector is influenced by all the factors discussed earlier as part of the overall corporate planning process. Underlying any choice is, by necessity, an appraisal of the risks versus the payoffs. The essence of past experience is that growth vectors within the existing market (or, at least, closely related markets) are much more likely to be successful than ventures into new markets. Therefore, diversification is the riskiest vector to follow, especially if it is attempted through internal development. The attractiveness of acquisitions for diversification is the chance of reducing the risks of failure by buying a known entity with (reasonably) predictable performance.

An equally crucial basic strategy choice is the degree of emphasis on innovation versus imitation. The risks of being an innovator are well known: few, if any, diversified corporations can afford to be innovators in each product market. There are compelling advantages to being first in the market if barriers to entry (through patent protection, capital requirements, control over distribution, and so on) can be erected, and if the product is difficult to copy or improve upon and the emerging phase is short.

The imitator, by contrast, can be put at a cost disadvantage by a successful innovator and must be prepared to invest heavily to build a strong market position. But even though profits over the life of the product may be lower, there are advantages to following the leader. The imitator's risks are much lower because the innovator has provided a full-scale market test that can be monitored to determine the probable growth in future sales. Also, the innovator may open significant opportunities to the imitator by not serving all segments or, more likely, by not implementing the introduction properly. Indeed, some managers believe that "being a close second" may be the optimum strategy.

The conscious decision to lead or follow pervades all aspects of the firm. R&D can have significantly different orientations to the market:

First to market—based on strong R&D, technical leadership, and risk taking.

Follow the leader—based on strong development resources and the ability to act quickly as the market starts its growth phase.

Applications engineering—based on product modifications to fit the needs of particular customers in mature markets.

Me-too—based on superior manufacturing efficiency and cost control.

Three fundamental questions then have to be asked of each new product or service being sought or considered:

○ How will a strong competitive advantage be obtained? The possibilities range from superiority in underlying technology or product quality to patent protection, to marketing requirements. Another dimension of this question is the specification of markets or competitors that should be avoided because they would blunt the pursuit of a competitive advantage.

○ What is the potential for synergy? Synergy, as Lee Adler notes, refers to "the mutually reinforcing impact a product market entry has on a firm's efficiency and effectivenss." It can be sought for defensive reasons, in order to supply a competence that the firm lacks or to spread the risks of a highly cyclical industry (the objective behind a number of mergers in the machine tool industry). Alternatively, synergy can utilize an existing competence such as a distribution system (notable examples are Gillette and Coca-Cola), a production capability, or promotional skills. In addition, Adler points out, "financial reinforcement may occur either because of the relative pattern of funds generation and demand . . . or because the combination is more attractive to the financial community than the pieces would be separately."*

○ What specific operating results are required? A company may focus, for example, on rate of market growth, payback period (despite its deficiencies, it is a reflection of the risk level), minimum sales level, or profit levels, cash flow, and return on assets.

There are many ways to summarize a checklist of questions for analysis; there are as many checklists as there are writers. Our preferred format is given in Table 3-1.

Checklists provide a disciplinary tool to ensure that no significant factor in an analysis is overlooked. However, they can often grow into a kind of company policy manual and ultimately a screening device. When checklists are used for screening, the final result is a weighted score derived from a worksheet of some sort. Overall ratings can easily conceal critical deficiencies in a market or business area that is buried under an "ability to commercialize" criterion and does not show up as importantly as it should.

_____ RESOURCES FOR FINANCING GROWTH

We have touched on the relationships among the profitability of a firm, its potential growth rate, and the cash flow to finance that growth. We can quantify these relationships intuitively as shown in Figure 3-5.

The box in the diagram represents the capital pool N employed by a firm; the fraction f_E of this pool is equity capital. The stream emerging from the side of this box rN represents total earnings (where r is the fractional return on capital). Part of this stream is used to pay interest charges I on the debt portion of the capital pool; i is the interest on the debt portion in the pool. The balance is divided, with the fraction f_D paid as dividends and the remainder $(1 - f_D)$ retained and reinvested. The retained earnings E_R are equity capital;

*_Harvard Business Review_ (November–December 1966), p. 59.

Table 3-1. Analysis of new products and expanded existing products.

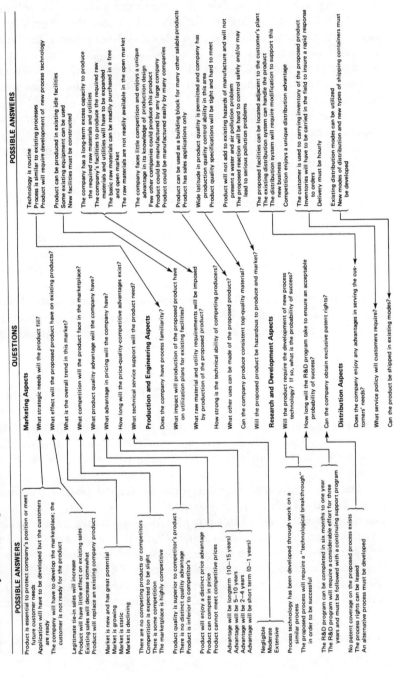

QUESTIONS

Marketing Aspects
What strategic needs will the product fill?
What effect will the proposed product have on existing products?
What is the overall trend in this market?
What competition will the product face in the marketplace?
What product quality advantage will the company have?
What advantage in pricing will the company have?
How long will the price-quality-competitive advantages exist?
What technical service support will the product need?

Production and Engineering Aspects
Does the company have process familiarity?
What impact will production of the proposed product have on utilization plans for existing facilities?
What raw material and utility requirements will be imposed by production of the proposed product?
How strong is the technical ability of competing producers?
What other uses can be made of the proposed product?
Can the company produce consistent top-quality material?
Will the proposed product be hazardous to produce and market?

Research and Development Aspects
Will the product require the development of new process technology? If so, what is the probability of success?
How long will the R&D program take to ensure an acceptable probability of success?
Can the company obtain exclusive patent rights?

Distribution Aspects
Does the company enjoy any advantages in serving the customers' needs?
What service policy will customers require?
Can the product be shipped in existing modes?

POSSIBLE ANSWERS (left)

Product is essential to protect company's position or meet future customer needs
Application will have to be developed but the customers are ready
The company will have to develop the marketplace; the customer is not ready for the product

Legitimate tie-in sales will increase
Product will have little effect on existing sales
Existing sales will decrease somewhat
Product will replace an existing company product

Market is new and has great potential
Market is growing
Market is static
Market is declining

There are no competing products or competitors
Competition is expected to be slight
There is some competition
The marketplace is highly competitive

Product quality is superior to competitor's product
There is no distinct quality advantage
Product is inferior to competitor's

Product will enjoy a distinct price advantage
Product can compete in price
Product cannot meet competitive prices

Advantage will be longterm (10–15 years)
Advantage will be 5–10 years
Advantage will be 2–4 years
Advantage will be short term (0–1 years)

Negligible
Moderate
Extensive

Process technology has been developed through work on a similar process
The proposed process will require a "technological breakthrough" in order to be successful

The R&D program can be completed in six months to one year
The R&D program will require a considerable effort for three years and must be followed with a continuing support program

No patent coverage on the proposed process exists
The process rights can be leased
An alternative process must be developed

POSSIBLE ANSWERS (right)

Technology is routine
Process is similar to existing processes
Product will require development of new process technology

Product can be produced in existing idle facilities
Some existing equipment can be used
New facilities must be constructed

The company has a long-term excess capacity to produce the required raw materials and utilities
The company's facilities to produce the required raw materials and utilities will have to be expanded
The basic raw materials can be readily purchased in a free and open market
The raw materials are not readily available in the open market

The company faces little competition and enjoys a unique advantage in its knowledge of production design
Few other companies could produce this product
Product could be manufactured by any large company
Product could be manufactured easily by many companies

Product can be used as a building block for many other salable products
Product has sales applications only

Wide latitude in product quality is permitted and company has production quality control ability in this area
Product quality specifications will be tight and hard to meet

Product will not add to existing hazards of manufacture and will not present a water and air pollution problem
The proposed reaction will be hard to control safely and/or may lead to serious pollution problems

The proposed facilities can be located adjacent to the customer's plant
The existing distribution system can handle the product
The distribution system will require modification to support this new business

Competition enjoys a unique distribution advantage

The customer is used to carrying inventory of the proposed product
Inventories will have to be carried in the field to insure a rapid response to orders
Delivery must be hourly

Existing distribution modes can be utilized
New modes of distribution and new types of shipping containers must be developed

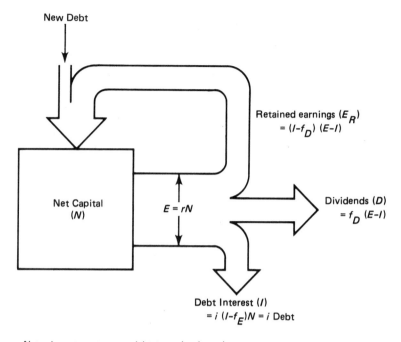

New Debt

Retained earnings (E_R)
$= (1-f_D) (E-I)$

Net Capital
(N)

$E = rN$

Dividends (D)
$= f_D (E-I)$

Debt Interest (I)
$= i (1-f_E)N = i$ Debt

Net reinvestment = new debt + retained earnings
Growth = net reinvestment/net capital
At constant debt/equity ratio:
Total reinvestment = E_R/f_E

Figure 3-5.　Cash resources for growth

reinvesting them adds to the equity component of the pool. To keep the equity fraction of the pool at some target level, the reinvested earnings must be balanced by an amount of new debt that makes the composition of the reinvested earnings stream entering the top of the box the same as the composition of the capital pool. Total reinvestment is equal to the retained earnings divided by the equity fraction f_E. Depreciation is not shown because it does not affect growth. The gain in capital when depreciation is reinvested just cancels the loss in capital when depreciation is first deducted. (This ignores the impact of inflation, discussed in Chapter 6.)

Consider a new manufacturing corporation, just starting in business. It needs money for plant, equipment, and "working" or operating capital. The money comes from selling stock and from borrowing from a bank. The former is called equity and the latter debt (long-term debt if it doesn't have to be paid back in less than one year). The total is the corporation's invested capital plus working capital. As the business grows, profits are reinvested (for example, to buy more equipment); they become part of the stockholders' equity. More important, the only way a manufacturing business *can* grow is by generating profits for reinvestment. The profits come from the volume of sales and a sales price that is sufficiently higher than all costs. Conceivably, a no-profit company could grow by selling more and more stock, or borrowing more, but in reality such additional funds would not be forthcoming. Even for profitable companies, the amount of additional funds from selling new stock is generally a very small fraction of the total.

Growth depends on the size of the new reinvestment stream entering the top of the box relative to the size of the capital pool. Growth rate and the cash required to finance growth can be related to the return on investment, the divident payout fraction, and the equity fraction by combining the equations for the various cash flows shown in the figure. The growth equation is

$$G = \frac{(1 - f_D)[r - (i)(1 - f_E)]}{f_E}$$

where G = capital growth rate
f_D = fraction dividend payout
f_E = fraction equity in capital pool
i = interest on debt portion in capital pool
r = fractional return on capital

The rate of return required to sustain any desired capital growth rate has been calculated for dividend payouts of 40 and 60 percent and for equity fractions of 50 to 70 percent. The results are shown in Figure 3-6. (Historically, although there have been variations in the debt/equity ratio for U.S. manufacturing companies, the average debt has been about 25 percent of the capital pool.)

To maintain the 9 percent per year average capital growth the U.S. chemicals industry had in the 1960s, at an interest rate of 4 percent and a 60 percent dividend payout the return on net capital would have to be about 16 percent. A 7 percent rate of return, more typical of some recent years, supports capital growth of only 3.3 percent per year. Dropping the dividend payout to 40 percent and even increasing new debt additions enough to reduce the equity fraction

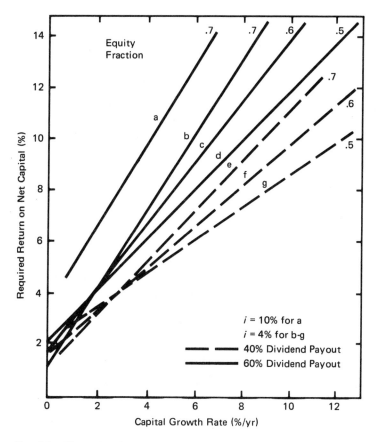

Figure 3-6.   Financing future growth

to 50 percent lowers the required return to 9.4 percent—still higher than the recent return of the industry.

Dropping the dividend payout and increasing the debt burden (if feasible) may be necessary to finance future growth. Another approach is to increase the supply of internally generated capital by raising the return on net capital. A 1 percent increase in return on net capital will permit a .6 percent per year increase in growth rate at a dividend payout of 60 percent and an equity fraction of 65 percent, and double this increase in growth rate—1.2 percent per year—at a dividend payout of 40 percent and an equity fraction of 50 percent.

But note that if interest rates are 10 percent per year, the attainable capital growth rates will be lower.

As noted above, a manufacturing company grows by investing more and more dollars in capital assets, preferably from retained profits. If that is not enough, the company can borrow money. As we will see later, the cost of debt is much lower than the cost of equity (that is, of issuing more stock). Interest paid on debt is a cost of business and is not subject to federal income tax, which is close to 50 percent for most large corporations. Therefore, to the borrowing corporation the real cost of 8 percent interest on debt is only about 4 percent.

Debt can take two primary forms: a loan from a bank or insurance company or a bond issued by the corporation. There is a limit to the amount of debt borrowing a corporation can undertake; the recent average for manufacturing companies has been about 25 percent of the capital pool (debt plus equity). For ranges up to 20 percent management may decide how much borrowing the company will do. Above that percentage, the banks and other lending institutions decide. If a company wants to borrow too much, it may find that the interest rate is prohibitively high, or the bank may just say no. If the company attempted to issue bonds, there would be no market for them.

Another way to get more dollars for capital investment is to cut dividends. This approach may be feasible for small companies, because the public will continue to buy their stock in the hope that company earnings will grow. Investors can then profit by selling their stock at a higher price later on. But it is a fact of life that large corporations must pay dividends of about half their after-tax profits (earnings) in order to sustain demand for their stock. If they do not, the price of their stock will drop—and the primary aim of every corporation president is to keep the price of common stock as high as possible. Paying dividends is one way. Another way is to give the public cause to believe that the company will grow in the future—that sales, profits, and investments will continue to expand.

The best way to raise profits (for increased investment) is simply to raise prices. This is often difficult in a competitive industry like chemicals. Since industrywide agreements on price levels or market shares are illegal under antitrust laws, each company must hope that all the others understand the simple principles of business economics.

Japan provides an interesting contrast with the U.S. system. There the government has worked cooperatively with business to encourage industrywide agreements when necessary—measures that would be illegal in the United States. The system is designed to insure that major corporations survive economic vicissitudes. Any analysis of the relationship between book value measures and market price must assume (1) steady-state conditions, (2) "rational" price actions in the stock market, and (3) full knowledge by investors of the company's steady-state future. In addition, the analysis draws heavily on the theory developed for making a choice between debt and equity in terms of the confrontation between the Modigliani–Miller hypothesis and "more traditional" hypotheses.

The percentage return to stockholders who invest in a company can be expressed as a combination of yield and growth:

Return to stockholders (%) = yield + growth

$$\text{Yield } (\%) = \frac{\text{beginning-year dividend}}{\text{beginning-year share price}} \times 100$$

Growth = growth in dividends (%/yr)

Yield, of course, can be expressed as a price/dividend ratio; then

Return to stockholders (%)

$$= \frac{100 \times \text{beginning-year dividend}}{\text{beginning-year share price}} + \text{growth in dividends } (\%/\text{yr})$$

In Figure 3-7 the return to stockholders has been related to the growth in dividends for different price/dividend ratios.

This protectiveness of the Japanese government greatly reduces business risk. Corporate bonds seem almost as safe as government bonds—because the government works to make them safe. Therefore, debt/capital ratios of Japanese manufacturing corporations have ranged up to 80 percent in recent times, compared with ratios of below 30 percent in this country.

_____ EQUITY MARKET VALUE OF A COMPANY

Increasing the market value of the equity owner's security is, correctly, a principal management objective of nearly all publicly owned corporations. Equity market value and the price/earnings ratio are recognized to be related to corporate earnings growth in some nebulous fashion. Unfortunately, some corporate managements frequently do not understand the quantitative relationships that exist and, worse yet, sometimes

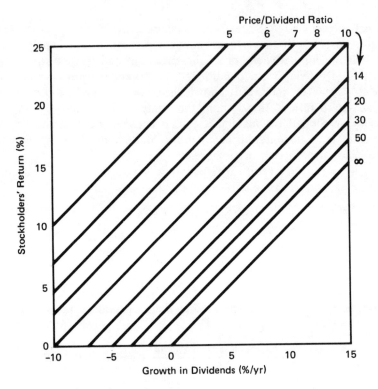

Figure 3-7. Stockholders' return

fail to recognize (or forget) that a desired long-term earnings growth rate can be achieved only by investing in projects (including R&D) that consistently meet a specified minimum rate of return. Measures of corporate performance that must interrelate include:

Price per share as established in the stock market
Corporate book measures as reported to the public
Internal cash flow measures used in planning the corporation's overall activities and evaluating specific projects (including R&D).

Our discussion here will be limited to the relationship between the stock market and corporate book measures; in Chapter 6 we will explore the relationships between the profitability of a specific project and corporate book measures.

If we assume a specific return to stockholders of say 10 percent, we can relate the price/earnings ratio to any combination of two of the following factors:

Dividend payout
Return on company's net worth
Growth in dividends

The interrelationships are shown in Figure 3-8. Remember we are interested here not in any particular corporate return but in the return to stockholders. (The equation used to generate the figure is given in the Notes at the end of the book.)

A price/earnings ratio of 10 can be maintained with any combination of two of the following parameters:

Figure 3.8   Price/earnings ratio, assuming reference to stockholder of 10 percent. ○ Net return on company's net worth—equivalent to stockholders' equity (%). ◇ Growth in dividends (%/yr).

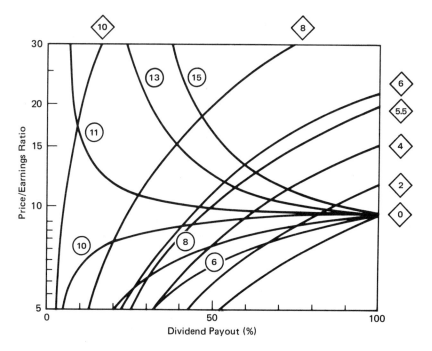

Dividend payout	50%
Return on company's net worth	11%
Growth in dividends	5.5%/yr

If one of the above parameters is not as indicated, the return to the stockholder will not be 10 percent.

Our discussion could be extended to consider other, perhaps more realistic situations that would be of interest to corporate management. The conclusion, however, would remain the same. Although the rate of return on new investments required to achieve the specified earnings growth rate can be calculated, earnings growth in the long run is the *dependent* variable and reflects the rate of return actually achieved on new investments. It may be desirable and is often necessary to have a target earnings growth rate and therefore a required rate of return on new investments. However, it must be recognized that if investment opportunities that meet the required rate of return cannot be found (or if investments are made that yield less than the minimum required rate of return), ultimately the earnings growth rate must decline. Under these conditions the stockholders' return and the market price for the stock will decrease. The desired earnings growth rate can be maintained only by investing in projects that equal or exceed the minimum required rate of return for new investments. This is true regardless of the form of investment, including R&D.

To summarize the concepts developed in this chapter: Assume an investor buys a share of common stock for $100. A look at the company's financial report shows it had earnings (profits after taxes) over the past 12 months of $10 for every share of stock. This means the investor bought at a price/earnings ratio of 10. The dividend payout is 50 percent or $5 per share.

Why would an investor pay $100 for a security that yields only $5 when he could have bought a risk-free bond that yields $8 per year? The answer is that he is thinking of the future. He knows the company management is committed to increasing earnings (by increasing the investment). If he has reason to believe that earnings will double in seven years' growth (a rate of 10%/year), the price of his share of stock will likewise double—providing the company can achieve the growth from retained earnings and debt rather than from issuing new stock.

More likely, if, because of the company's past record and announcements of investments or discoveries, a number of investors really believe the stock price may double (that is, if the "image" of the company is highly favorable), the price/earnings ratio will anticipate the happy events to come. It may increase almost immediately to, say, $13. Now the share of stock is worth $130. The investor may sell it for a short-term gain of $30 or hold it as a long-term investment.

If, on the other hand, some months after the investing public has trustingly expressed its opinion that company earnings will grow at a rate of 10 percent per year, some unexpected unfavorable development occurs and it appears that next year's profit may drop to $8 per share, disillusionment will quickly set in. Investors will lose interest in the stock, and the price/earnings ratio may drop to 6. Now the share of stock can be sold only for $60.

This is the way it is in the stock market. Today's stock prices are totally dependent on expectations of tomorrow's profits. The track record of the company and the credibility of its management in talking about the future are all-important.

The same thinking applies to the price offered by one company when it wants to acquire another company. Experts from the acquiring company can estimate the value of factories, equipment, inventories, and the like. The experts combine their estimates with the record of the last few years' earnings. They estimate the future growth of those earnings and from that decide on a price/earnings multiple (5, 10, 15, and so on) for the whole company (or each share of stock of the company), which establishes what it is worth as an acquisition. Then, because the acquiring company will be buying most or all of the stock (or the equivalent) of that company at once, it must add a premium of, say, 20 to 40 percent to rebalance the supply–demand relationship of the acquired company's stock.

chapter 4

The Basic Economic Structure of a Business

Any meaningful analysis of R&D requires an understanding of the fundamental accounting concepts, principles, and methods that are critical to the financial management of a company. Our framework for discussion will be limited to the concepts of cash flow, funds flow, and income flow and their relationship to a company's balance sheet. Understanding how accounting data are used for management decisions is critical to the development of R&D activities that support a company's present business activity and provide for growth. The attitude of R&D managers should be: How can the profits of the company be protected and enlarged? There would seem to be three essential requirements:

An awareness of profits.
An ability to quantify predicted profits.
An understanding of what constitutes adequate profits.

Obviously, the income received from the sale of commodities produced by a company contributes to profits. But it is often not realized that every physical action taken by a company reduces profits. Therefore, an organization must strive constantly to minimize the cost of necessary actions and to eliminate as many unnecessary actions as possible.

Quantification of predicted profits is not just a matter of mechanics. It is also a matter of definition. Depending on the definition of "profit," the same set of figures may or may not show that a profit exists. Therefore, profit can be defined only in terms of the mechanics employed to create a profit statement. It is essential that the mechanics of quantifying profit be fully understood and agreed on in advance.

FINANCIAL ANALYSIS

Understanding what constitutes an adequate profit is most important of all. After a profit-making opportunity has been recognized, and after the predicted profit has been quantified according to an agreed-on definition, the question remains: Is this enough profit? If this question cannot be answered by those who have recognized and quantified the profit, many profit-making opportunities will be judged inadequate and will never be brought to the attention of those who think otherwise. Let us put these concepts into perspective.

When a business is first established, there usually exist only (1) a supply of capital and (2) a plan for spending it profitably. In the case of a corporation the capital, in the form of money, comes from stockholders, who expect to share in the profits of the spending plan, and from bondholders and mortgagors, who expect payments of interest and principal "off the top" before profits are given to the stockholders. The stockholders take the largest risk, because the return of their money depends on the success of the spending plan. The bondholders and mortgagors take a lesser risk, because their money can usually be recovered through the forced sale of the physical assets that are created under the spending plan. Let us look, then, at the spending plan (perhaps it should be called a profit plan) from the point of view of the stockholders. Where will their money go, and where will their profit come from? These questions can be answered by examining a balance sheet and a statement of income and retained earnings, documents that everyone has seen in companies' annual reports.

A balance sheet presents the total result of the expenditures of money from the first day a corporation exists until the day the balance sheet is prepared. Rather than make up a fictional company, we have used published financial information of E.I. du Pont de Nemours & Company, a large R&D-oriented organization with an outstanding record of accomplishments. Obviously, the results of a financial analysis will depend on the time period selected, but the techniques are always the same. If the original stockholders of Du Pont could have looked ahead to December 31, 1975, Figure 4-1 is approximately what they would have seen.

The right-hand side of this balance sheet shows that the original stockholders, together with those who came later, have supplied $773

FIGURE 4-1. Consolidated balance sheets, E.I. du Pont de Nemours & Company and Consolidated Subsidiaries (in millions of dollars)

Assets		Liabilities	
Cash and marketable securities	139	Notes and loans payable	540
Notes and accounts receivable	1,163	Accounts payable and accrued liabilities	400
Inventories	1,220	Income taxes payable	171
Prepaid taxes and other expenses	51	Long-term debts	887
Investments in nonconsolidated affiliates	103	Other liabilities	228
Property plant and equipment, at cost less depreciation and depletion	3,592	Deferred income tax	126
Deferred charges and assets	100	Deferred investment tax credit	152
Goodwill, patents, and trademarks	56	Equity of minority shareholders in consolidated subsidiaries	83
		Shareholders' equity	
		Capital	773
		Earnings reinvested	3,062
	6,425		6,425

Note: Since the numbers have been rounded off, the columns do not add up exactly.

million to Du Pont. Du Pont also borrowed substantially, with long-term and short-term debts that now stand at $1,427 million. (This debt is in the form of notes, loans, mortgages, and bond issues.) In addition, Du Pont owes various bills amounting to $400 million. The total liability (including certain income tax and other accrued-liabilities accounts) down to this point is $2,590 million.

What has been done with this money? The left-hand side of the balance sheet shows that $3,592 million is invested in fixed assets—land, buildings, and equipment that Du Pont owns. (Not shown are many millions of dollars in fixed assets that have previously been bought, used, and discarded.) The fixed assets that the corporation presently owns cost $8,585 million and have depreciated to the value of only $3,592 million. Depreciation does not necessarily reflect physical obsolescence or decline in worth. Rather, it is an accounting statement of the amount by which past incomes have covered the cost of fixed assets while being utilized for other purposes as well. (We will come back to this in later discussions.)

The left-hand side of the balance sheet shows other assets. Du Pont has a bank account and owns certificates of deposit and other securities with a total value of $139 million. It owns $1,220 million in chemicals, fuel, and other commodities that are in inventory in tanks, warehouses, and other places. Customers owe Du Pont $1,163 million for goods that have been shipped to them—"receivables" that are entirely collectable in cash. Du Pont owns part of several other companies throughout the world; the value of its share is conservatively estimated at $103 million. Patents and trademarks are considered by some impossible to value; they are valued at cost or a nominal figure to insure against overstatement. Goodwill arises when another concern is acquired and payment for its assets is larger than their fair market value. Du Pont has valued its patents, trademarks, and goodwill at $56 million. This completes (along with some prepaid taxes and deferred charges) the list of assets totaling $6,425 million.

The question immediately arises: If Du Pont has assets worth $6,425 million and if its debts total only $2,590 million, what about the remaining $3,835 million? The answer is it is the share of the assets owned by the stockholders, including the $773 million share they advanced. Therefore, the $3,835 million labeled "shareholders' equity" is not really a liability in the common sense; it is a statement of the stockholders' equity in a "going concern," and when it is placed on the liabilities side of the balance sheet, the two sides balance.

An important observation is now in order. The holders of 50

million shares of stock (common and preferred) have an equity amounting to the $15.46 per share they supplied to the corporation plus $61.24 per share in reinvested earnings. This is approximately $77 per share. Why, then, did Du Pont stock sell for about 1.5 times as much on the Stock Exchange during 1975? The answer is only partly found in the balance sheet. According to generally accepted standards in the business world, the capital structure of Du Pont is evidence of strength and solidity. That is to say, the liquidation of the corporation is practically inconceivable. Du Pont maintains the ability to do the thing it was created to do: supply chemicals and other products at a profit.

Remember that the assets of Du Pont are far greater than the amount of money supplied by stockholders and lenders. This situation could have come to pass only if Du Pont reinvested some of the profits generated in previous years. Here we have evidence of an essential characteristic of a business: the ability to pull itself up by its own bootstraps; the ability to increase its own worth. Moreover, it is necessary that this ability actually be exercised. If it is not exercised, the business will either die or be killed by other businesses that compete for the same market.

How does a business exercise its ability to grow? Where does the extra money come from? These are questions that a prospective stockholder raises and questions that a statement of income will help to answer. If the original stockholders of Du Pont could have looked ahead to December 31, 1975, they would have seen Figure 4-2.

The statement of income and retained earnings shows that in 1975 Du Pont had the ability to sell $7,222 million in chemicals and polymer (and receive $56 million in other revenues). The statement also shows that Du Pont paid a substantial price to do this. The labor and materials that went into the production and distribution of commodities cost $5,410 million. The costs of advertising, selling, research studies, engineering and personnel to guide the business totaled $709 million. The fixed assets shown on the balance sheet were partially recovered by depreciation allowances (see Chapter 7) amounting to $580 million; $126 million was paid as interest to those who had loaned the corporation the $1,427 million shown on the balance sheet. Remaining was $453 million, of which various governments took $176 million for taxes. This left $277 million in "profit."

A profit of $277 million in one year! For this, stockholders have purchased 50 million shares of stock, an investment that has permitted the creation of almost $7 billion in assets. Was it worth the effort

Income	
Sales	7,222
Other revenues	56
Deductions	
Cost of goods sold	5,410
Selling, general, and administrative expenses	709
Depreciation, amortization, and obsolescence	580
Interest	126
Subtotal deductions	6,825
Net income before income taxes	453
Income and other taxes	176
Net income after income taxes	277
Dividends paid to stockholders and minority interests	220
Additions to reinvested earnings	57
Shareholders' equity at beginning of 1975	3,753
Shareholders' equity at end of 1975*	3,835

*Stockholders' equity at the end of 1975 increased by $82 million. Of this, $25 million came from the sale of capital stock; the balance, from additions to reinvested earnings.

FIGURE 4-2. Consolidated statement of income and shareholders' equity, E. I. du Pont de Nemours and Company and Consolidated Subsidiaries (in millions of dollars)

and the gamble? In 1975 stockholders received 80 percent of the profit as dividends. This is $4.25 for each share of stock, which sold on the Stock Exchange for upward of $100 per share. On the face of it, 1975 Du Pont stock may not look like the best investment for a stockholder, especially a new stockholder. Most banks pay a return higher than 4.25 percent, with no gamble at all except for a national disaster, which would also affect Du Pont.

There is clearly more here than meets the eye. Du Pont must offer an enticement that banks do not. That enticement is simply the ability to produce income at a higher rate in the future. It is no coincidence that an objective of Du Pont is "to attain maximum long-term *growth* in earnings per share." Du Pont's balance sheet attests to a solid foundation for this growth. The statement of income attests to an abundance of opportunities for small percentage changes that could add up to substantial increases in profits. And there is one other important factor: diversification. Du Pont insures against financial disaster by producing a number of different products for different uses—that is, by keeping all its eggs out of one basket.

At the same time, many of Du Pont's activities offer unique opportunities for growth. It is extremely unlikely that Du Pont will be in the position of the buggy-whip maker before the automobile was invented—who also probably had an excellent balance sheet and an excellent statement of earnings.

Table 4-1 summarizes some relevant Du Pont financial data over the period 1969–1975. The numbers look large, but they must be viewed from the perspective of the capital requirements of Du Pont over the next decade. They also show some significant changes in just a few years.

Of special note is the final ratio, return on total capital. Companies and investment analysts use this type of ratio as a guide for making comparisons among individual operating divisions (or with other companies), for promoting successful operating managers, and for selecting new projects. Thus it is difficult to compare a low-debt company with a high-debt company, unless we add interest (paid on debt) to profit to eliminate the effect of purely financial manipulations. Indeed, this is exactly what knowledgeable analysts do in order to put companies on the same operational basis, treating all as though they had no debt. If dollars paid as interest were added to pretax profit, they would be subject to an income tax (federal and state) of about 50 percent. Therefore, only *half* the interest should be added to net earnings to give an adjusted earnings figure. This makes a significant change in net income, the numerator of the return on capital fraction.

Knowledgeable analysts also make a change in total capital, the denominator of the fraction. In the case of Du Pont, depreciation and depreciated fixed assets (valued at $3,592 million) are artificial terms for income tax purposes (see Chapter 7). Thus analysts would use $8,585 (the original cost of the assets) as the value and add the difference between the two numbers to the $6,425 million to get a new "gross total assets," or total utilized investment (see pages 77–78). They would then use this figure in the denominator of the return on capital fraction.

Another of many possible adjustments takes cognizance of the fact that profits cover the entire year, whereas the total assets in the balance sheet refer to the last day of the year only. Therefore it may be more reasonable to average the asset value for the end of the year ($6,425 million) with that for the beginning of the year ($5,980 million).

Shortly after Irving S. Shapiro took over as chairman of Du Pont

TABLE 4-1a   Du Pont financial data (in millions of dollars)

Year	Total Assets (A)	Common Stockholders' Equity* (B)	Long-Term Debt (C)	Notes and Loans (D)	Total Debt (E)	Interest Charges (F)	Net Income (G)	Net Income Before Interest Charges (H)
1975	6,425	3,596	887	540	1,427	126	277	403
1974	5,980	3,514	793	320	1,113	62	404	466
1973	5,052	3,355	250	169	419	35	586	621
1972	4,284	3,029	241	—	241	24	414	438
1971	3,998	2,856	236	—	236	18	357	375
1970	3,740	2,725	162	—	162	18	334	352
1969	3,452	2,615	147	—	147	15	369	384

*Start of year.

TABLE 4-1b   Du Pont financial ratios

Year	B/A = Fractional Stockholders' Equity	C/B = Long-Term Debt/Equity Ratio	F/E = Pretax Interest (%)	H/F = Interest Coverage	H/A = Return on Total Capital Before Interest Charges (%)
1975	.56	.25	8.8	3.2	6.2
1974	.59	.22	5.5	7.5	7.8
1973	.66	.07	8.3	17.7	12.2
1972	.71	.08	10.	18.2	10.2
1971	.71	.08	7.6	20.8	9.4
1970	.73	.06	11.1	19.5	9.4
1969	.76	.06	10.2	25.6	11.1

in 1974, he surprised many in the financial community and the chemicals industry by announcing that Du Pont, which became a giant on such new products as nylon and Orlon, would pull in its innovative horns. With existing businesses demanding more and more capital, Du Pont would cut back by 75 to 80 percent the number of new ventures it would pursue. Since then, the company has cut back even more than its chairman predicted—from about two dozen major new ventures to just two. Among the projects dropped entirely were polypropylene packaging film, a carpet underlining, and a barrier resin for plastic bottles.

Normally, such a drastic retreat would show up as a sharp decline in the research and development budget. But Du Pont, which had sales in 1976 of $8.4 billion, spent $353 million on R&D, which is just a bit more than it spent in 1974. Thus, despite its retreat from new products, Du Pont continues to support one of the largest R&D budgets in U.S. private industry. A major aim of Du Pont's R&D today: to find solutions to urgent problems of raw materials and production that could undermine the company's ability to compete in the 1980s and beyond.

Concern for its feedstock base is new at Du Pont, and so is concern for the supply of other raw materials. Some of the company's big competitors—Dow Chemical and Union Carbide—have based many profitable ventures on their strong market position in key building-block chemicals such as chlorine, benzene, and ethylene. But Du Pont in the past has been content to buy those commodities, relying on its heavy R&D commitment to turn them into products that have high added value—from nylon to its current "hot" fiber, Kevlar.

That strategy has served Du Pont well, but it failed seriously in 1974. With the increases in oil prices, the company's costs for feedstocks and other raw materials soared by 250 percent. Other factors too—competition and inflation—were forcing a change in Du Pont's R&D game plan. The company was adept at introducing a new product, aggressively marketing it, and continually improving it. By the time a competitor caught up and prices declined, Du Pont's profits on its investment were usually assured. But even Du Pont took a beating when the recession of 1974–1975 hit a synthetic fibers industry afflicted with marked overcapacity. A big factor also was the expiration of some key Du Pont patents.

The industry was unable to raise prices to keep up with escalating costs of materials and energy. Synthetic fibers, which accounted for 35 percent of Du Pont's sales in 1974, produced $126 million in profits, well off the pace of the previous two years. And the following year fiber profits collapsed to a disastrous $6 million.

"The old era died at the close of 1973," says the company's 1975 annual report. "The gun was loaded by the oil embargo . . . and the trigger was pulled by double-digit inflation."

Today Du Pont is putting more weight than ever before on research and development of new processes that will squeeze cost out of existing products. It has developed, for example, a new way to make adiponitrile—an intermediate product used in the manufacture of nylon—that cuts the chemical's cost by 6¢ per pound. When the company completes two big replacement plants in 1981, the new process will cut costs by $60 million a year. The push is visible in Du Pont's other businesses too. Company scientists recently found a way to reduce the silver content of photographic film by 10 percent without affecting the quality or sensitivity of the image. Last year the new process saved the photo products department $7.5 million.

The financial consequences of some of these events are evident in Table 4-1. Du Pont is finding that back-integration into raw materials is "expensive"; and the nature of Du Pont's interest in R&D is subtly changing. Du Pont is retreating in part to the strategy of selectivity that it pursued in the late 1940s and early 1950s.

OPPORTUNITY DIFFERENTIAL ANALYSIS

Every action taken by a company is either a profit-making or a profit-losing "opportunity." Thus the economic analysis of an opportunity should be portrayed from the viewpoint of the company. It is not sufficient to circumscribe the analysis with the arbitrary limits of a production unit, a plant location, a marketing area, or even a division. Nor is it sufficient to limit the analysis to an immediate period of time. The analysis should embrace all the consequences an action will have in other areas and other times as well.

Theoretically, every economic analysis should present the predicted balance sheets and statements of income and retained earnings for a company as a consequence of undertaking an opportunity. This is impracticable, for the reason that no one can know all the future actions that a company will take. Fortunately, it is not necessary to make an analysis of this magnitude except, possibly, when a new major business venture is contemplated. Prediction of the future can be limited to those corporate areas where the future will depend on the decision at hand.

The aim of every economic analysis is, therefore, to answer the question: What difference will this course of action make to the financial future of the company? Hence every economic analysis can be restricted to analyzing the *differential* between proceeding with the proposed action and not proceeding. But note that "not proceeding" is a course of action in itself and can present a large number of alternatives. There are many ways that a company might not proceed. One alternative is to continue along the course of action that was being followed before the proposal was made. Other alternatives include variations on the proposed action. Thus, in almost every instance, an economic analysis will comprise two kinds of differentials. One, often called (in self-contradiction) an *absolute differential*, determines if *any* positive action of the kind proposed is justified. The other, often called a *comparison differential*, determines which of the possible variations will constitute *the* proposed action.

An absolute differential presupposes that a company will continue along some course of action if the proposed action is not taken; although what that course may be is often difficult to define. In future years, many decisions will be made, and many business modifications will take place. Not all these modifications will lie outside the economic area of the proposed action. For example, technological changes may reduce the requirements for a given raw material. A proposed action to reduce the cost of that raw material must certainly take cognizance of the decreasing demand. Even the best alternative method of reducing the cost would be of no avail if changes in other areas eliminated the need for the raw material.

Similarly, technological changes will be implemented by expenditures, and the economic analysis of them must take into account the alternative of reducing the raw materials cost. Therefore, many actions are mutually dependent and mutually exclusive. In creating an absolute differential, the analyst faces the task of predicting what future actions will be taken if a proposed action is not adopted, and to what extent the action will modify them. Only in this way can the analyst be sure that the preferred course of action, determined by comparison differentials, is better than leaving things as they would otherwise be.

In choosing the alternatives to be considered, the analyst must define clearly the decision that will be guided by the analysis. For example, a decision to construct a new production unit authorizes not only the expenditures required for assets but also the expenditure of operating funds. A "yes" decision to construct is also a "yes" decision to operate. Each of these subactions should be analyzed separately, by comparison differentials, to insure that the best alternatives are chosen and that the overall course of action is optimized. No analysis should be considered complete until reasonable study has been given the possibility of more profitable means of accomplishing the same objective.

The order of execution of the two kinds of differentials is almost always optional. Most often, an approximately absolute differential (a "rough appraisal") is prepared, utilizing the alternative action that intuitively seems best. If this absolute differential is not attractive, selection of the best alternative action by means of comparison differentials is not usually warranted; the proposal is abandoned. When an absolute differential is attractive, further investigation by comparative differentials is made. In the end, the decision to act or not to act is based on the absolute differential. This alone comprises the justification for the preferred course of action.

The relative ease of execution of the two kinds of differentials will, of course, vary. In general, however, comparison differentials are easier to construct than absolute differentials. Comparison differentials are most often concerned with alternatives that have several economic elements in common. If these common elements are identical in magnitude and timing, they can be disregarded.

For example, the decision to produce a certain quantity of a new product may involve a choice among different processes. One or more comparison differentials will determine which process is best, without regard to the ultimate disposition of the product. However, if the product is to be employed "captively" in subsequent manufacture of other products, determinations of the absolute differential presents a problem: What is the income? Unless the income can be stated, the profit cannot be calculated. But there is no income, because the captive product is not sold. In this situation, which occurs more often than not, there are several possible methods of analysis:

○ A fictitious selling price can be employed. If the captive product is actually being sold by some other company, the fictitious selling price can be based on a market price. Otherwise, it is necessary to estimate what the price would be if the product were sold.

○ The income from the sale of the derived products can be used. In order to retain qualitative similarity between the income element and the cost elements, the production cost of the derived products must also be included in the analysis. Inasmuch as the production cost of the derived products includes the cost of producing the captive product, the justification for producing the former is the justification for producing the latter.

○ The analysis can omit the absolute differential. This is the least acceptable choice but the only one in a large number of instances. For example, if the captive product is 600 psi steam, there may be no market price and the derived products constitute practically all the production at the site. The omission of the absolute differential, and with it the economic justification for building a steam plant of any kind, is equivalent to saying: "Our analysis tells us that this steam is needed to produce increased quantities of many products. Each of these products has been previously justified, and each justification has included the cost of steam from a proposed new plant. Thus the products have collectively justified the construction of a new steam plant whenever they need it. They will need it *x* months from now, and construction should begin without further delay." A similar kind of "justification by association" is employed for utilities and many auxiliary departments.

A question is frequently asked about the analysis of opportunities: What degree of accuracy is required? The answer is enough to permit the correct decision to be made. This answer appears to sidestep the question, but no other simple explanation can be given. It is the decision maker who decides, and the analyst must be aware of the kind of decision that will be made. This may range from a decision to proceed with a more detailed analysis to a decision to spend many millions of dollars. The duty of the analyst is to inform the decision maker of the time and methods necessary to make an analysis with a given degree of accuracy. Together, they can then decide what degree of accuracy is required.

TYPES OF OPPORTUNITIES

There are four monetary categories that identify the activities of a company: income, deductions (costs), assets, and liabilities. Of these categories, all except liabilities are important in the analysis of opportunities. This is not to say that corporate liabilities are unaffected by the courses of action that a company takes. At present, however, there is no simple approach for predicting the effect of an action on liabilities. Liabilities are related to the capital structure and financing decisions of the company. They involve separate optimization decisions that are the concern of the treasurer and top management.

Opportunities may in turn be classified into three groups according to their impact on the monetary categories: income increase, cost reduction, and new asset reduction or postponement.

INCOME INCREASE   There are several ways to increase the amount of corporate income:

Sell a product that the company has not sold before.
Sell increased volumes of existing products.
Raise the price of existing products.
Sell knowledge in the form of patents and technological experience.
Sell unused assets.

Of these, the first two account for the great majority of opportunities to increase income.

Increasing income *by selling a new product* is the first choice of a growing business. In fact, it is the only choice if a high corporate growth rate is to be maintained. Sooner or later older products reach a profit plateau because of competition and market saturation. But new products grow rapidly only if they have been properly designed for the intended market.

The sale of new products has its disadvantages, however. The research and development costs are not insignificant. The rule of thumb attributed to Du Pont is that the cost of taking a "new venture" product through early market development is ten times the initial R&D outlay; the cost of full-scale commercialization is ten times again as much. Expensive new production facilities must usually be provided. And increased costs for labor, raw materials, utilities, advertising, selling, and management will be incurred. The prediction and evaluation of the incomes, asset expenditures, increased costs associated with new products is one of the most important and active areas of economic analysis in a company.

Inherent in the analysis of a new product, or for that matter any product, is the interrelationship among volume, selling price, and cost. As volume increases, the cost per unit volume of product decreases. At the same time, a progressive increase in volume may depend on a progressive decrease in the selling price. There can be some volume beyond which the decrease in cost does not keep pace with the decrease in price, with a resulting decrease in total profits. This volume is sometimes referred to as a "point of diminishing returns."

When the price–volume relationship is known accurately, the optimum volume can be calculated, because the cost–volume relationship is nearly always predictable. Unfortunately, the relationship between price and volume is rarely known with accuracy. This circumstance gives rise to much speculation about price and volume, so that the analysis of a large number of alternatives may be required in order to understand the sensitivity of profit to price–volume changes and provide a reasonable basis for decision.

Another way to increase income is *to sell increased volumes of existing products*. A firm like Du Pont sells a large number of products that are costly to develop and manufacture. It would be a waste if sales remained below the capacity of these facilities. Moreover, an increase in production to the limit of existing capacity would not require significant increases in assets or labor costs. The last increment of production is, in most cases, the least expensive per unit of volume. Thus it is partly to protect the existing volume level and partly to take advantage of volume economies that a concentrated effort is made to increase the volume of existing products.

In the study of volume expansions within existing facilities, the greatest need for economic analysis is not for facilities that operate continuously to produce a specified product; rather, it is for continous or batchwise operations that can produce several products alternately. As these products grow in volume, they compete for use of the facility until one or more are forced into new facilities. The question then is: For which products will new facilities be built? In general, the answer is the product with the greatest predicted growth. But this is not always the case. It may be that the predicted growth can be attained only at a selling price insufficient to justify the expenditures for new facilities. The determination of the proper course of action in these cases will be the subject of many economic analyses. One possible course of action is to forgo the potential increase in income.

Volume expansion of products that are made continuously in specially designed facilities may also require analysis, since sales efforts often result in predicted volumes that exceed existing capacity. Prediction and analysis of the additional income, expenditures for new assets, and additional costs are similar to those for a new product.

Raising the price of existing products is often the most difficult way to increase income. It can and has caused a reduction in volume, which may well outweigh the gain achieved by raising the price. However, a decline in volume even at higher prices rarely lures competitive sellers into the marketplace; competition is lured primarily by perceived opportunities for growth.

Selling knowledge (such as technology or business know-how) has apparently been lucrative for some firms. Because knowledge cannot be valued precisely, its sales price has been a matter for judgment and negotiation by management. There is good reason to believe that the going price for knowledge is low. When the sale of knowledge may also induce an increase in the supply of a product, the analysis is complicated, since it must consider the impact that buyers of the technology will have on existing markets.

Sales of unuseful assets is relatively new for most firms. In view of rapidly changing technology and premature obsolescence of products, a constant stream of usable, but unuseful assets is being retired from service. Many of these are versatile enough to serve some other function; others are not. A profitable market for them may exist. In addition, smaller plant locations can be consolidated into larger, more profitable combinations, thereby making the small plant salable "in situ." Since such a sale would provide a potential competitor with a low-cost, ready-built facility, this opportunity requires a very detailed analysis.

COST REDUCTION   Cost and deduction are not the same. Indeed, the word "deduction" as it appears on a statement of income and retained earnings (see Figure 4-2) is not universally understood. To the Internal Revenue Service, a deduction is an amount by which income is shielded against taxation. To the manager of a production unit, money spent is not a deduction but an expense. And to the accountant, money spent for anything but assets is a noncapital expenditure.

We will use the word "cost" for any expenditure of money other than for assets. Therefore, costs will include the expenses of operating production units as well as the expenses of operating all other departments such as laboratories and offices, wherever they may be found. Like the definition of deduction on a statement of income and retained earnings, cost does not include income taxes, which are not really expenditures in the sense of directly supporting the growth of a company. In contrast to the definition of deduction, cost does not include depreciation, which is not an expenditure to begin with.

The term "cost reduction" is widely misinterpreted for reasons other than the types of costs included. Misinterpretation stems primarily from the fact that a cost can be expressed as a dollar amount or as dollars per unit of product. These two ways of expressing cost come in conflict most often when the volume of an existing product is increased. Although an expansion in volume increases the total dollar cost, it usually brings about a decrease in the average cost per unit of product. Thus it is often felt that a cost reduction has taken place.

It is true that the unit cost has been reduced, and this reduction has an important bearing on pricing decisions, but it must be remembered that every economic analysis is a differential comparison between the dollars in one alternative and the dollars in another alternative. Therefore, the term "cost reduction" should not be used in economic analyses unless there is an absolute reduction in the dollar amount of one or more costs.

There are innumerable ways to reduce costs. Whatever method is used, the reduction will have an impact on the company's assets. Cost reductions can be grouped into two major classes, depending on whether they increase or decrease assets.

Cost reductions that require an increase in assets are by far the larger in number. The reason is that costs represent either labor or material. As long as no change is made in the function for which the labor or material is purchased, costs can be decreased only by reducing the amount of labor or material. Often such a reduction can be accomplished only by installing new time-saving or material-saving equipment. Thus there is an increase in the fixed assets of the corporation.

At the same time, however, there is a decrease in one of the current assets, inventory. Since inventory is nearly always valued at the cost of producing whatever volume exists in stock, reduction in the cost per unit of production (as opposed to other costs such as selling and research) will result in a decrease in the current assets of the corporation. The decrease in current assets may, in a rare case, be larger than the increase in fixed assets. (However, there is a fundamental difference between current assets and fixed assets, which we will discuss later in the chapter.)

A common example of a cost reduction that leads to an increase in fixed assets and a decrease in current assets is the decision to produce a commodity instead of purchasing it. This situation is called a *make-versus-buy* analysis. In it, the disadvantage of expending funds on fixed assets is compared, through a comparison differential, with the advantages of lowered costs through making a commodity and a lower investment in the inventory of that commodity. However, whether the decision is to make or to buy, an absolute differential must also be prepared to determine if the chosen alternative is preferable to having no commodity at all.

In this regard, it is pertinent to note that a sure way to reduce costs is never to choose either alternative. This, also, is an alternative. It has no added costs. And no added profits. Hence it poses the danger of too much cost reduction.

Another cost reduction variation is an *own-versus-lease* analysis, in which the disadvantage of expending funds on fixed assets is compared, by means of a comparison differential, with the advantage of lowered costs through owning. Important to the decision is an intangible disadvantage in the lease alternative: contractual requirements for lease payments may restrict a company's freedom of choice in the future.

Facilities that may be leased—such as automobiles, tank cars, barges, and warehouses—can usually be sold. Ownership is not necessarily as irrevocable a commitment of funds as leasing. As with a make-versus-buy analysis, an own-versus-lease analysis is not complete until an absolute differential has shown that the preferred alternative is itself a profitable opportunity.

Closely related to leasing is the contracting of services that a company could perform with its own personnel. Examples of such services are engineering, maintenance, research, computing, protection, and even economic analysis. Expenditures for assets in the "own" alternative may or may not be a significant factor in the analysis. Very often, the numerical analysis will rest primarily on the simple comparison of costs. However, the intangible factors of speed of action, secrecy, safety, dependability, accuracy, and versatility for other possible applications must not be overlooked.

Cost reductions that permit a decrease in assets have increased in number but not in popularity in recent years. They differ from the reductions just discussed in that both the current assets and the fixed assets of the company are decreased. There is only one way to achieve such a reduction: by abandoning part or all of the function for which the assets were previously utilized. Hence the economic analysis of this type of cost reduction is frequently called *abandonment analysis*.

Abandonment analysis involves no comparison differentials. The course of action proposed is to cease the operating expenditures for a function and to relinquish the assets utilized by that function. This action is either taken or not taken. There are no other choices.

The impetus to abandon a function may arise from at least three circumstances:

○ The function is not contributing a profit. This circumstance does not usually call for analysis. The course of action is obvious, except when contracts exact a penalty for abandoning the function, when the contribution to income is obscure, or when a company is consciously forgoing current income to build a base for the future.

○ The function is contributing a small profit, but other functions could utilize many of the same assets to produce a larger profit. Moreover, other functions may have to purchase new assets if the assets being utilized by the existing function are not made available. Thus the economic analysis rests on the attractiveness of the new assets for alternative use. In effect, the existing function must justify its continued existence against the new assets that its abandonment would assist.

○ The function is contributing a small profit, but other functions could utilize the same people to produce a larger profit. (Note the similarity to the point above.) In a real sense, people are assets, even though they are not reported as such on a balance sheet. No economic analysis is possible here, because the value of people cannot be quantified. However, it is necessary to remain alert to wastage of "the people asset."

NEW ASSET REDUCTION OR POSTPONEMENT   Perhaps it is incorrect to call a reduction or postponement of new assets an opportunity. An opportunity should increase the amount of profits, and assets are a means to that end. But even though assets are necessary, certain types such as fixed assets may be looked on as a necessary evil. By nature they are closely related to costs. In this light, therefore, there is merit in seeking to reduce or postpone expenditures for new assets, provided that the long-range profits are not disproportionately lowered.

The reduction or postponement of new assets can come about only by increasing income or reducing costs. There must be a reason for having a new asset in the first place, before an opportunity to reduce or postpone the asset can occur. In reality, therefore, opportunities of this kind are alternatives to the cost reduction opportunities just discussed.

The most common ways to increase income are to sell new products and to increase the sale of existing products. Both approaches result in an increase in fixed assets and current assets. However, current assets may also be decreased if cash is used up to generate initial sales. Seldom will the demand for a new product, or the increased demand for an existing product, require the immediate operation of a new facility at full capacity. Depending on the process, there is usually an optimum capacity for a new facility below which the costs per pound of product are higher and above which the facility becomes unwieldy to operate. If the additional sales amount predicted for some future year is a multiple of the optimum capacity of facility, there is no point in building more than one facility immediately. The remainder can be postponed, leaving the funds available for more immediate opportunities.

Much depends on the definition of optimum capacity. No two people will agree on what the optimum capacity is quantitatively. However, there can be no quarrel with the general philosophy that the optimum capacity is the one that results in the greatest profit over the long run. In some cases, it may be preferable to build a facility of less than optimum size or to convert an older facility in order to retain funds for other opportunities. The penalty here is higher costs per unit of volume, but this may be worthwhile if the funds released for other opportunities produce a counterbalancing profit.

If possible, the initial facility should be designed so that it can be expanded to optimum size at an early date. This, no doubt, will ultimately result in a higher total cost and higher interim costs than if the facility has been built optimally in the beginning. Still, the postponement opportunity should not be overlooked. Shortly we will see that the discounted cash flow method is especially adapted to this type of economic analysis.

In a somewhat similar connection, we should note that the philosophy of reduction and postponement extends downward from the facility as a whole to its smallest components. In the design of any facility, each component should be examined for its immediate need and its long-range cost of operation. The money saved by purchasing a slightly smaller component may well counterbalance a consequent increase in operating cost. Certain components provided in the name of safety or versatility may not be mandatory or desirable. In every facility, there are manifold opportunities for the profitable reduction of asset expenditures.

Thus far our discussion of asset reduction has been aimed primarily at fixed assets. However, the potential for reducing current assets, especially inventories, should not be overlooked. Inventory is usually valued by multiplying the volume in stock by the full production cost per unit of volume. An inventory can be considered as layers of production, each one having a different full cost depending on when it was produced. An increase in production quantity, or a decrease in the cost of production, will reduce the cost per unit of volume and thus the value of inventory.

The value of inventory can also be decreased by a simple reduction in the volume maintained in stock. This is not always easy to do, since minimum inventory levels are needed to give good service to customers. If a company does not have enough inventory to fill a sales order, that order, and all future orders from the customer, may be placed with a competitor. But the potential for asset reduction through decreases in the volume of inventory exists in infinite variety. It is always a profitable avenue of exploration.

A very common opportunity to reduce assets deserves special mention here. This opportunity arises whenever a product is produced in batches (even if the process is continuous) rather than continuously. In many instances, the production of batches in series cuts down on the need for equipment cleaning or reduces the amount of off-specification products. The capacity of the equipment to produce on-specification products is thus increased.

Some reduction in costs can come about through learning over time. As operating personnel become familiar with the process, major improvements can be made in the utilization of raw materials and manpower. In addition, raw materials can be purchased in larger, more economic quantities. However, the volume of product in inventory may also be increasing at a rapid rate.

It is not unusual for the capacity of a "batch" facility to far exceed the sales demand. The preferred course of action *may* then be to produce a run of several batches, shut down while the accumulated inventory is sold, and then begin another run. The analysis of optimum run length is complex. It is especially difficult when the facilities can be employed for other products between runs. The operating costs, inventory levels, production rates, interest costs, and sales demands of all the products must be embraced by the analysis.

Stockpiling is a rather peculiar type of asset postponement opportunity and often has far-reaching consequences. Stockpiling occurs when a predicted increase in the sales volume of an existing product can be more than matched by the existing capacity, but only temporarily. Sooner or later an increase in fixed assets will be required, but it can be postponed by operating in excess of demand while demand remains low.

A significant increase in the inventory asset will result. The value of inventory reaches a peak when the demand equals the increased rate of production. Thereafter, inventories decrease to normal as volume is withdrawn from stock and sold. Meanwhile, investment in a new facility is postponed. The object of the economic analysis is to weigh the advantages of postponing the investment in the fixed assets against the disadvantages of increasing the current assets temporarily.

Often there are complicating factors. The increased volume of inventory may require the purchase or lease of additional storage facilities. The additional inventory value will be subject to property taxation and must be insured. The production costs of the additional inventory do not qualify as deductions against income taxes while the inventory exists. The raw materials for the product may be in short supply because they are needed for other products. There is also the risk of obsolescence of materials. And if two or more production facilities exist, the increased production must be allocated among them by a careful consideration of the differences in production costs. The complete economic analysis of a stockpiling opportunity can be very complex.

A final observation: inasmuch as a cost reduction opportunity causes an increase in one or more assets, the decision not to make a cost reduction results in a reduction of assets. Thus, for every cost reduction opportunity that is not elected, there is a counterpart asset reduction opportunity. Striking the ideal balance requires considerable ingenuity.

CRITERIA OF BUSINESS GROWTH

Our earlier discussion of the economic structure of Du Pont (see Table 4-1) brought out the fact that in 1975 the company earned $398 million in net income (after taxes but before interest charges) through using $6,425 million in assets. This is a 6.2 percent after-tax return on the total capital employed. Is a 6.2 percent return sufficient to Du Pont?

There is no exact answer to that question. If a 6.2 percent return is equal to or better than the return achieved by any business similar to Du Pont, it might be termed acceptable, if not sufficient. If Du Pont has the ability to produce a higher return, 6.2 percent is not even acceptable. Du Pont's after-tax return over the last ten years has exceeded 10 percent. The percentage difference is larger than it seems: a 10 percent return in 1975 rather than the actual 6.2 percent would have netted the company an additional $244 million in profits after taxes. Based on past performance, there is reason to believe that 10 percent is not the best Du Pont can do.

One approach to developing criteria about acceptable ranges of

return is to look at long-term rates of return for key investment media:
U.S. common stocks, foreign equities, long-term (20-year) high-quality
corporate bonds, and Treasury bills or other short-term instruments.
From 1926 through 1975 U.S. common stocks (Standard & Poor's
500 stock index) provided a return of 9 percent per year with all
dividends reinvested. Over this same period comparable investments
in long-term, high-quality corporate bonds and Treasury bills returned
3.8 percent and 2.3 percent respectively; the inflation rate was also
2.3 percent. For the past 103 years realized returns for stocks, bonds,
and bills respectively were 9.4 percent, 3.9 percent, and 3.6 percent
per year, and the inflation rate averaged 1.9 percent per year.

Looking to the future, we can make some interesting rate of
return forecasts on the basis of present bond prices. Long-term
corporate bonds have recently yielded in the range of 8 to 9 percent
per year, undoubtedly because investors anticipate much higher rates
of inflation than those experienced on average over the past 50 to
100 years. With high-quality bonds yielding between 8 and 9 percent,
is it reasonable to expect stocks—with their much greater uncertainty
of return—to yield only 9 percent or so in total return to investors?

In fact, over the past 50 years stockholders have received a 5.2
percent per year premium for holding stocks rather than bonds; over
103 years the premium has been 5.5 percent. It seems much more
plausible for forecasting purposes to expect the average spread or
risk premium between long-term bonds and stocks to remain roughly
in line with historical experience. Carrying this logic to its conclusion
in the current market environment, we can estimate total returns
of 13 to 14 percent for U.S. common stocks over the next 10 to
20 years.

Recent studies of the impact of inflation on nominal and real
rates of return seem to support these crude forecasts. Professor John
Lintner of the Harvard Business School has suggested that the adverse
effects of an increase in expected future rates of inflation or an
increase in unanticipated "transient" inflation are concentrated in
the transition period from one rate of inflation to a higher rate. Thus,
if the consensus of economic forecasters proves correct, and if inflation
does stabilize at a more moderate rate after the 1973–1974 runup,
future nominal equity returns will incorporate the higher inflation
rate. Lintner goes on to show that "*after* the period during which
inflation rates have increased, even a continuance of higher (instead
of lower) constant steady-state (and fully anticipated!) rates of inflation
will leave the mean *expected real* rate of return on holding equities

just as high as it was before the increase in inflation rates, and expected nominal returns will be proportionately higher.''*

It should be noted, however, that there are more pessimistic opinions about corporate stock prices and earnings. Pessimism is evident in the fact that over recent years stock prices have declined even though corporate earnings have increased along with the investment. The Dow Jones average of stock prices (of a group of the nation's major corporations) exceeded 1,000 in 1965 and dropped below 800 by January 1978. Obviously, investors have been fearful about future real growth in earnings. They fear that inflation will have a negative effect on profits (costs will rise faster than prices); they worry that the government and the public are losing appreciation of the profit motive and the free-market economic system; they are afraid that restrictive laws and regulations are putting the government in control. That has been the pervasive view in recent years, as evidenced by low price earnings ratios (Du Pont 11, Dow Chemical 8, Union Carbide 7, and Monsanto 7 in early 1978).

If our estimates of desired (or possible) stockholders' rate of return are correct, Du Pont must significantly increase *its* corporate rate of return. For the 1975 corporate return levels, we would estimate from Table 4-1 a price/earnings ratio of 10 rather than the 20 or so the market has placed on Du Pont.

Perhaps our criterion is too harsh; it is certainly after the fact. Before developing others, we should define the significance of any criterion. Fundamentally, a criterion is a standard of performance. But what do we mean by "performance"? Performance can be expressed by the dollar amount of any profit element in a given year, or it can be expressed by the relationship between profit and the assets utilized to produce it. The former measure has little meaning, because a business with $1 billion in assets cannot be expected to produce the same amount of income as a business with $2 billion in assets. As long as the assets are changing (and presumably growing), a criterion must take cognizance of the amount of assets. Since debt financing is ultimately limited, money for the expansion of assets must eventually come from income.

Furthermore, increased income is derived primarily from expanded assets. If all asset expansion is derived from income, the percentage of income to assets is also the percentage by which the income could grow. As long as this rate of income generation is maintained, growth

---

* "Inflation and Security Returns," *Journal of Finance* (May 1975).

is limited only by the number and magnitude of available opportunities. Therefore, a criterion is a measure not only of absolute monetary performance but also of the rate at which performance should improve. Improvement of performance is itself a kind of performance, and indeed the most important.

Performance can be measured by the relationship between profits and assets. This relationship is most conveniently expressed as a ratio of one year's profit to the assets utilized in that year. The 6.2 percent return before interest charges for Du Pont in 1975 is an example of this sort of measurement, which judges performance after the fact. We are concerned primarily with prediction, and a criterion must also be adaptable to an expression of predicted long-range performance.

Several criteria are available for expressing expected performance *before* the fact. One is the ratio of profit to assets in one particular future year. Another is the ratio of the average profits to the average assets utilized in several future years. A third is the algebraic sum of the total profits in several future years and the total money spent for assets in those years. Obviously, this sum should be positive.

In setting the value of a criterion, we must exercise great caution. If the criterion is set too high, few plans of action will qualify, and there will be little growth. If the criterion is set too low, ill-conceived actions will be taken, resulting in profits that are not commensurate with the resources used to create them. People must have a high but obtainable goal to spur them to do their best.

Moreover, the risk factor must be taken into account. Some plans of action are riskier than others, and a criterion set high enough to compensate for exceptional risks would be too high for cases of little risk. The subject of risk will be discussed in Chapter 9; it is introduced here because of its relevancy to setting criteria. Note that risk exists only before the execution of a plan of action, never afterward. There is no doubt, on Sunday morning, about the outcome of Saturday's game.

Thus a criterion must have two applications. First, it must measure the sufficiency of an action after it is taken. Second, it must insure beforehand that a proposed action will be sufficient once the action is taken and the risks have exacted their toll. Inasmuch as we are concerned with predicting the results of a proposed action, it is evident that the second application is more pertinent.

Another consideration in setting a criterion is the physical size of a business as measured by its assets. As noted previously, a business with $1 billion in assets cannot be expected to produce the same amount of income as a business with $2 billion in assets. However, there is reason to believe that the *rate* of growth of the smaller business will usually exceed that of the larger business. As a business grows in size (and usually also in age), income must increase geometrically for a constant growth rate to be maintained. Such an increase is difficult to sustain in the short run and impossible in the long run. It is true that the resources of the larger business offer opportunities of greater size and diversity than those available to the smaller business. But in the final analysis opportunities come from the ideas of employees, and employees do not increase at a constant geometric rate. Therefore, a criterion must take the size (and the age) of the business into account.

Finally, a criterion should not necessarily have a constant and unchanging value. From time to time, it should be revised to reflect (1) the availability of opportunities and (2) the availability of money. If there are more opportunities than the supply of money can exploit, the criterion could be raised to eliminate the least desirable opportunities. If there is more money than there are "acceptable" opportunities to spend it, the criterion could be lowered to make more opportunities acceptable; or, better yet, the dividend payout to stockholders could be increased. Essentially, the criterion should be adjusted to bring the opportunities and the money supply in balance.

Much has been written on this subject, and there is wide agreement that *money is not generally in short supply* (although it may be for some firms). The limiting resource is good ideas that lead to acceptable opportunities. *Money can always be obtained, at a price, for an acceptable opportunity.* Obviously, the lower limit of acceptability bears some relationship to the "price" of money, but the criterion must be set higher than that to insure profit and growth. How much higher a criterion needs to be, or can be, is of great concern to all companies.

In the remaining part of this chapter we will consider alternative measures of business performance. Each measure emphasizes a specific aspect of growth or cost that should be reviewed periodically.

COST OF CAPITAL   In economics, interest refers to the cost of using capital. The accounting definition of interest that we have been using differs from an economic definition in two respects:

Interest refers only to the charge for using debt capital; accountants do not record a charge for using equity capital.

Interest is not an element of cost in the sense that labor and materials are so treated; rather, interest on debt capital is regarded as an expense.

Many people have argued that accounting should adopt the concept of interest used in economics.

Table 4-2 shows what happens if interest on the use of both debt and equity capital at Du Pont is accounted for as a cost item. The cost of capital is

$$\frac{\$290 \text{ million}}{\$6,425 \text{ million}} \times 100 = 4.5\%$$

In these calculations the interest is on an after-tax basis. Therefore, the interest on borrowing is reduced by 40 percent, Du Pont's effective tax rate. The 6.2 percent return before interest charges for Du Pont appears insufficient when viewed against the total cost of capital, which is a lower boundary for a company's rate of return. This is particularly true if our forecast of the impact of inflation on future nominal rates of return is anywhere correct.

Cost of debt is easy to calculate. The cost of equity capital is a difficult concept. It should be higher than the interest rate on debt (more than 8 percent early in 1978) by a margin that allows for the increased risk. (The cost of equity capital applies to retained earnings as well as to new stock offerings, since such earnings could be used by the company or the stockholders for purposes other than reinvestment.)

There are a number of mathematical derivations of cost of capital; some are based on dividends, others on earnings. (See Chapter 12.) A few years ago, when top-grade corporate bonds paid 8 percent interest, one estimate of the cost of equity capital was 10 percent,

Table 4-2. Calculation of 1975 Du Pont annual interest rate (in millions of dollars).

	Interest Amount	Interest or Dividend Cost	After-Tax Rate (%)
Interest on borrowings	1,429	75	5.3
Preferred stock	239	10	4.2
Common shareholders' equity	3,596	205	5.7
Nonmonetary liabilities	1,161	—	
	6,425	290	

as indicated in this example of the cost of (total) capital for a hypothetical corporation:

	Cost Before Tax	Cost After Tax	Market Value (millions)	Weighted Cost
Short-term debt	8.5%	4.4%	5	.55%
Long-term debt	8.	4.2	10	1.05
Common stock	10.	10.	25	6.25
Average cost of capital				7.85%

Note the reduction of the cost of debt (interest) for federal income tax not paid on interest.

How can Du Pont "get away with" paying only 40 percent income tax? The answer for most large corporations can be ascertained only by close analysis of financial figures. For one thing, the numbers in annual reports to stockholders are not the same as the figures reported to the IRS. For federal purposes corporations may use maximum depreciation rates to lower their tax; in the reports to stockholders they use straight-line (average) rates. Also, taxes on profits of foreign operations often can be deferred.

RETURN ON INVESTMENT   For many years specified values for a quantity known as return on investment (ROI) have been used as criteria of a company's performance. We define the ROI arbitrarily by the following ratio:

$$\frac{\text{Income less all costs, income taxes, depreciation}}{\text{Current assets + fixed assets at cost}}$$

More specifically, the ROI equals:

From Statement of Income		*From Balance Sheet*
Income		Cash and Securities
Less cost of goods sold, selling, general and administrative expenses	DIVIDED BY	Plus receivables
		Plus inventories, property, buildings, and equipment
Less depreciation		valued at cost.
Less income tax		

The numerator of this fraction is variously abbreviated net income after taxes (a misnomer, because interest charges have not been deducted) and ROI income. The denominator is called total utilized investment (TUI).

Expressed as a percentage, the ROI criterion usually specified in the literature is 10 to 15 percent. This contrasts with the actual Du Pont ROI of only 3.6 percent in 1975. (Do not confuse this after-tax ROI with the 6.2 percent after-tax return previously referred to. The investment bases are quite different: an $11 billion base cost versus a $6.2 billion base cost after deductions for depreciation and obsolescence.)

Obviously, Du Pont is falling far short of the 10 to 15 percent ROI criterion. But this does not necessarily mean that the criterion is too high (although it may be). It does mean that a 10 to 15 percent ROI criterion has a built-in safety factor for risk. Judging by the demonstrated performance of 3.6 percent ROI (and assuming that Du Pont has grown through courses of action that have predicted, on the average, ROIs of 10 to 15 percent), it would appear that many of the suspected adversities have actually manifested themselves, at least during 1975. But what if the profit from a proposed action were guaranteed? What percentage ROI would justify taking what riskless action? This remains to be seen.

At this point, it may be worthwhile to recall that we have already encountered two definitions of profit: net income after taxes on net investment (net on net) and ROI income. In both cases, the income is before interest payments. It is worth noting, before yet another concept of profit is introduced, that clear understanding of definitions and precise usage are essential to the communication of common objectives. Choose the terms that best suit a particular economic analysis, define them, and stick to them.

The best measure of *adequate* ROI is sufficient profit (after payment of dividends) to enable the company, through reinvestment, to grow as fast as other companies in the same line of business. In recent years this figure has been about 7 to 8 percent per year for leading companies in the chemical business. For Du Pont in 1975, the amount needed to sustain this growth rate would be .08 × $11 billion, or about 1.2 billion. (The $11 billion is total assets plus the depreciation reserve, as discussed above.) Approximately $580 million would come from depreciation (written as a cost but available for reinvestment); the remaining $620 million would have to come from profits after dividends. Obviously, Du Pont fell far short of this figure in its 1975 profits. The company could borrow almost $300 million ($1.2 billion × .25) without increasing its debt/equity ratio (increasing that ratio is undesirable for the ordinary growth of a business); but even then Du Pont would fall far short of its previous growth trend:

Depreciation	$580 million
Potential added debt	300
Reinvested profit available for growth*	57
	$937 million

*$277 million − $220 million in dividends

Obviously Du Pont could have cut its dividends below $220 million for 1975, but this would have drastically affected its image to the investing public—and the price of a share of Du Pont stock. (Investors would have feared the cut in dividends might extend to future years.) Instead, by maintaining the dividends, Du Pont projected the impression that 1975 was merely a one-time bad year.

RETURN ON SALES   Another quantity that is often employed as a criterion is return on sales (ROS), which is arbitrarily defined as a ratio:

ROI income ÷ net sales

"Net sales" means essentially the same thing as "revenue" shown on a statement of income and retained earnings (except for deletion of dividend and interest income, if any). The word "net" means that the income has been adjusted to reflect products returned by customers. Inasmuch as the dollars received from customers constitute by far the largest source of income, the word "sales" is apt. However, the primary reason for using "net sales" instead of "income" is to prevent confusion of the terms "income" and "net income." The word "sales" is graphic and admits of no misinterpretation.

Like the ROI, the ROS is expressed as a percentage. In 1975 Du Pont had an ROS of 5.6 percent. As a criterion the ROS is useful, but not for the same reason that the ROI is useful. The ROI is a criterion of asset productivity. The ROS is a criterion of *potential*. It quantifies the sensitivity of the ROI to possible changes in profit.

With Du Pont's 1975 costs of operation over 94 percent of income, it is easy to see that a small change in the remaining 6 percent could significantly change the ROI. A simple example, rounding off Du Pont financial data, should make this clear. In 1975 Du Pont supported $100 million in sales with a TUI of about $152 million. Each $100 million in sales cost over $94 million, with an ROI therefore of 3.6 percent. If the costs could be reduced to about $90 million (a reduction of about 5 percent), Du Pont's ROI would increase to 6.5 percent, a rise of 50 percent. Therefore, the ROS gives a "feel" for the consequences of changes in income and costs.

Although the ROS is widely used as a measure of profitability, it is much less desirable than the ROI for capital-intensive companies like those in chemicals. For noncapital-intensive companies such as those in cosmetics, paint, retailing, and services, the ROS is, by default, a good available measure. Return on capital employed (debt plus equity) would be just as good.

While no value of the ROS has been established as a criterion of profitability, it is a useful financial ratio for management control. *Key Business Ratios*, a Dun & Bradstreet publication, provides comparative ratio data for many industry groups. For the chemicals and allied products industry the average ROS in 1975 was 6.4 percent.

TURNOVER Turnover (TO) is arbitrarily defined as a ratio:

Net sales ÷ total utilized investment

Like the ROI and ROS, the TO is expressed as a percentage. Like the ROS too, it is a criterion of potential. It gives an idea of the degree of activity from which profit and growth will emerge.

Obviously, if net sales were only 50 percent of the total utilized investment (for a TO of 50 percent), there would be very little chance of generating enough income to produce a spectacular ROI (costs being as large as they are). In 1975 Du Pont had a TO of 66 percent. No value of the TO has been promulgated as a criterion, but chemicals and allied products generally have a TO ranging from 80 to 100 percent.

The product of return on sales and turnover equals the return on investment:

$$\frac{\text{ROI income}}{\text{Net sales}} \times \frac{\text{net sales}}{\text{TUI}} = \frac{\text{ROI income}}{\text{TUI}}$$

$$\text{ROS} \times \text{TO} = \text{ROI}$$

This often serves as a quick way to check arithmetic.

OPERATING RATIO Operating ratio is defined as follows:

Cash flow ÷ net sales

Cash flow, in turn, is defined as:

ROI income + depreciation

Operating ratio is expressed as a percentage and is closely related to return on sales. In 1975 Du Pont had an operating ratio of 13.5 percent, which was 7.9 percentage points higher than its ROS. This follows from the fact that the depreciation deduction was 7.9 percent of sales. Note that the operating ratio is free of association with the assets. Insofar as sales and costs can be altered without altering the assets, the operating ratio permits an assessment of cause and effect.

There is a deeper significance to the operating ratio. It implies a definition of depreciation different from the definition inherent in the statement of income and retained earnings. The subject of depreciation will be considered more fully in Chapter 7; suffice it to say here that the ROI and ROS criteria, and the statement of income and retained earnings on which they are based, regard depreciation as a genuine cost that must be deducted before profit is calculated. The operating ratio does not regard depreciation as a cost; this concept of depreciation is also employed by the discounted cash flow method (see Chapter 6).

chapter 5

The Appraisal
of
Alternatives

Maximization of economic performance as measured by profits is a well-accepted management goal. R&D management is faced with a continuing need to choose among alternatives when considering such activities as:

Development of a new product (new to the world or new to the firm)
Improvement of an existing product
Development of a new process
Improvement of an existing process
Technology-licensing decisions
Technical service activities

To make the choice, the manager must integrate information on markets, production costs, capital requirements, research needs, and other factors to establish some rational basis for a decision. The

Table 5-1. Factors influencing the profitability of an alternative.

Factor	Possible Accuracy of Forecast	Effect of Possible Error
Installed cost of fixed investment	Good	Intermediate
Working capital	Good	Intermediate
Construction period	Good	Minor
Initial startup expense	Fair	Intermediate
Sales volume forecast	Poor	Major
Product price forecast	Poor	Major
Cost stream over project life	Good (direct) Fair (overheads)	Major
Economic life	Poor	Intermediate to major
Depreciation life	Good	Intermediate
Depreciation method	Good	Intermediate
Salvage value	Poor	Minor
Income tax rate	Poor	Major
Inflation rate	Poor	Major
General business conditions	Fair	Intermediate to major
Exchange rate	Poor	Major
Cost of development	Fair	Minor to intermediate
Time to complete development	Fair to poor	Intermediate
Minimum acceptable rate of return	Policy choice	

basic steps in making the decision are clear; the execution, less so. The manager must estimate costs, estimate benefits, convert estimates into a measure of desirability, and compare the measure with a standard.

Decisions involving capital expenditures and those involving R&D expenditures pose some of the same problems. This chapter will focus on the similarities; differences will be discussed in Chapter 9.

To make a decision involving capital expenditures, a manager must decide what factors will influence the profitability of each alternative and to what extent. The number of important factors is greater than is usually believed. Moreover, the desired benefits occur in the future, and the future is never known with certainty. A partial classification of factors that influence profitability is given in Table 5-1. The manager is therefore faced with making a forecast of many factors over a period of time. The following discussion is limited to capital projects.

A cumulative cash position chart can be used in profitability analysis to portray the relationships among the alternatives under consideration. At any given time, a chart like the one in Figure 5-1 will show the cumulative cash impact that a project can have on a company. The chart includes most of the complexities of a normal project, except that equal annual investments and straight-line depreciation (see Chapter 7) are assumed for ease of illustration.

The first investment shown is for land at a single point before time 0. This is followed by a gradual increase of fixed investment during construction. Working capital is shown as being added instantaneously at time 0—obviously an approximation. Equal annual profits are assumed to start at time 0, bringing the cumulative cash position line upward at a constant slope throughout the life of the project.

Figure 5-1.   Cumulative cash position chart

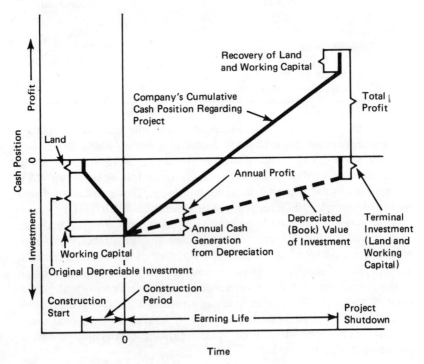

All these elements can be seen clearly, as can the construction period and economic life of the project. The value of the investment remaining on the company's accounting books is approximated by the dashed line: in this case, economic project life and depreciation life have been made identical.

Note that the cumulative cash position line is negative initially and that some time passes before it gets back to 0. This is a characteristic of all alternatives that have a front-end financial load, including capital investment and R&D expenditures. The size and timing of that front-end "hole" will be critical in determining the attractiveness of an alternative.

BREAKEVEN ANALYSIS

Once an alternative has been chosen, a breakeven chart can be used for analysis. A breakeven chart shows the sales volume below which there will be no ROI income, along with the impact of fixed costs, direct costs, and investment (through depreciation). It also provides an insight into the vulnerability of ROI income to possible changes in volume and price.

Fixed (period) costs are those associated with owning a unit (for example, depreciation, taxes, and insurance). They have little relationship to the production rate. Direct costs may be either variable or semivariable. Variable costs increase or decrease in direct proportion to the level of production. Chemicals, catalysts, and utilities fall into this category. Semivariable costs vary with the production rate, but not in direct proportion to it. Examples include supervisors and service people.

Depending on the project, labor and maintenance can be fixed, variable, or something in between. For a plastics film extruder that can run on one, two, or three shifts, labor costs are variable. For a continuous-flow process, labor is a fixed cost because the full workforce may be needed even at a low operating rate. (If the rate is too low—say, 30 percent capacity—the unit may be shut down for some months.)

There are several stages in the development of a breakeven chart. The annual period costs and depreciation allowances are plotted against the sales volume, which is assumed to be the same as the production volume. In the classical case horizontal straight lines are produced, as shown in Figure 5-2a. The horizontal lines show costs that are

Figure 5-2.   Development of a breakeven chart

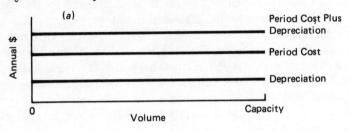

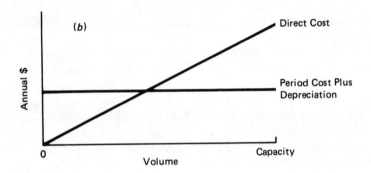

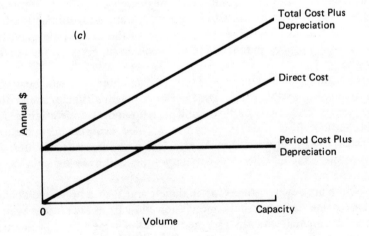

approximately constant over almost the entire range of plant operation. Depreciation is obviously one of these. So is plant overhead, such as an allocated fraction of the plant manager's salary, the first-aid room, and the plant cafeteria (assuming the project is one of many at a single plant location). Also included are salesforce expenses, supporting R&D expenses, and administrative (mostly clerical) costs.

The next step is to plot direct costs, as shown in Figure 5-2*b*. A straight line of constant slope originating at 0 volume is often assumed. The total cost plus depreciation is derived by adding the ordinates of the two curves, as shown in Figure 5-2*c*.

The final step is to plot the curve of annual sales. If the price is constant over all ranges of volume, the curve will be a straight line of constant slope originating at 0 volume. The vertical distance between the curves for sales and for total cost plus depreciation is the ROI income. At the intersection of these curves the ROI income is 0. The value corresponding to the intersection is called the breakeven volume or breakeven point, as shown in Figure 5-3*a*.

There are many ways to prepare breakeven charts. A common variation involves the stepwise addition of fixed cost (such as additional operating labor) as volume increases, as shown in Figure 5-3*b*. Such a chart might illustrate the economics of a unit that can operate on one, two, or three shifts. Maintenance may be as high when the equipment is not used as when it is.

A profitability index sometimes referred to in connection with the breakeven point, but only remotely related to it, is the breakeven price. It is the sales price that would produce sales income exactly equal to the total cost plus depreciation at *capacity volume*. In other words, it would produce an ROI income of 0 at capacity.

One disadvantage of a breakeven chart is that it does not show ROI income directly; the figure must be derived by subtraction. Also, if several possible sales curves are exhibited, the chart becomes visually confusing. The modification shown in Figure 5-4 is sometimes used to obviate these difficulties. Here period cost plus depreciation is plotted negatively. In a separate calculation, the direct cost is subtracted from the sales income to yield the marginal income, which is plotted along the vertical (y) axis.

The intersection of the marginal income curve with the horizontal (x) axis is the breakeven point, and the ordinate at any volume is the ROI income. This presentation, often called a cost–profit–volume or CPV chart, allows several marginal income curves representing different prices to be plotted clearly.

Figure 5-3. Breakeven chart showing (a) breakeven point, (b) stepwise variation

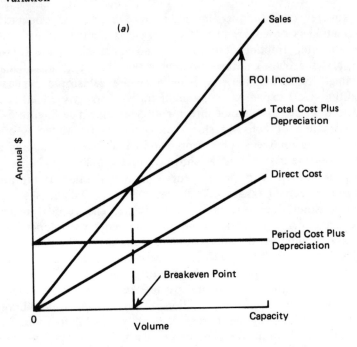

(a)

Sales

ROI Income

Total Cost Plus Depreciation

Direct Cost

Period Cost Plus Depreciation

Breakeven Point

Annual $

0

Volume

Capacity

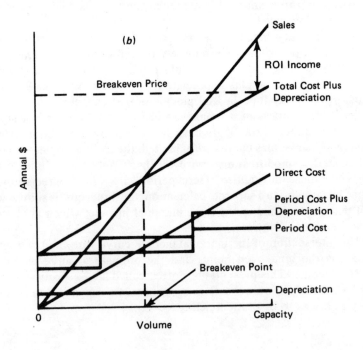

(b)

Sales

ROI Income

Breakeven Price

Total Cost Plus Depreciation

Direct Cost

Period Cost Plus Depreciation

Period Cost

Breakeven Point

Depreciation

Annual $

0

Volume

Capacity

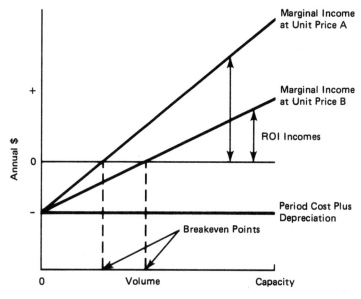

Figure 5-4. Cost-profit-volume chart

A potentially serious error can be introduced with a CPV chart, since it assumes that period cost and depreciation are constant over the full range of volume. In most cases they are not constant, and ROI income may be overstated at higher volumes while the apparent breakeven volume may be too low.

If the possibility of error is recognized, the CPV chart can be useful in optimizing the profit potential of a facility for which several products compete. Assuming that period cost and depreciation do not depend on the product being made or the volume produced, the marginal income curves for the individual products are plotted end to end out to the capacity of the facility. The optimum product mix for the facility can be determined by varying the price or the volume spanned by each product. A profit optimization chart is shown in Figure 5-5.

CRITERIA FOR DECISION MAKING

Cash position charts and breakeven charts present a mass of information. The following criteria are often used to summarize this information in a useful and understandable form:

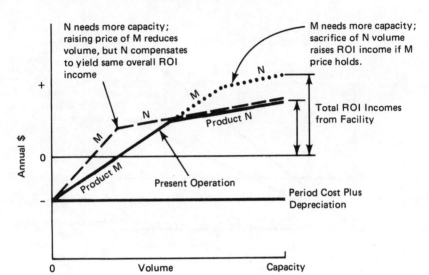

Figure 5-5. Profit optimization chart.

Payback (payout time)
Accounting rates of return (book return, net-on-net, ROI)
Benefit/cost ratios
Discounted cash flow (DCF) measures
○ Internal rate of return (variously referred to as yield, interest rate of return, discounted cash flow return, investor's method, marginal efficiency of capital)
○ Net present value (present worth)

Payback, accounting rates of return, and benefit/cost ratio calculations are done for a specific year. The discounted cash flow method projects all cash receipts and expenditures over a period of time, discounts the receipts and expenditures to reflect the time value of money, and expresses the result as a single dollar sum. In contrast to the other criteria, this method permits noncapital and capital expenditures to be combined. An example will help clarify how each criterion is used. Suppose we are considering four investment alternatives:

| | | Net Cash Proceeds | |
Investment	Initial Cost	Year 1	Year 2
A	$10,000	$10,000	$ —
B	10,000	10,000	1,100
C	10,000	3,762	7,762
D	10,000	5,762	5,762

In all cases, the initial cost is the capital investment. The net cash proceeds (cash flow) are depreciation (or the equivalent) plus profits after taxes. To keep calculations at a minimum, we will assume that all the alternatives are riskless (or of equal risk), all receipts and expenditures are at year end, and there is no salvage value.

Payback is the time required for the stream of cash proceeds produced by an investment to equal the original cash outlay required by the investment:

$10,000 Investment	Net Cash Proceeds		Payout Time in Years	Rank
	Year 1	*Year 2*		
A	$10,000	—	1.	1
B	10,000	1,100	1.	1
C	3,762	7,762	1.8	4
D	5,762	5,762	1.7	3

The payout time can be modified to include cost of capital.

Accounting rate of return expresses profitability as an annual percentage of the capital employed. There are several ways of computing it:

Average net income (after depreciation)

   ÷ average investment (i.e., average book value after depreciation)

  Average net income (after depreciation) ÷ original investment

Single-year net income (after depreciation)

     ÷ average, current-year, *or* original investment

   Cash flow (income + depreciation) ÷ original investment

Other measures, including ROI, may be used.

If we are working from financial data, many factors will influence the ROI, as shown in Figure 5-6. For our example, let us define ROI as:

     Average income ÷ average investment

We then obtain the following:

$10,000 Investment	Average Cash Inflow	Average Depreciation (Straight Line)	Average Income	Average Book Value	ROI (%)	Rank
A	$10,000	$10,000	—	$5,000	— %	4
B	5,550	5,000	550	5,000	11	3
C	5,762	5,000	762	5,000	15.2	1
D	5,762	5,000	762	5,000	15.2	1

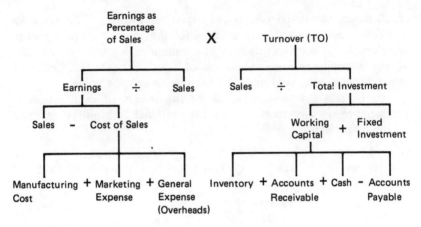

Figure 5-6.   Return on investment

Note that the rank order based on ROI is different from that based on payback. The criterion we choose has influenced our ranking and could influence our decisions.

Benefit/cost ratios have their origin in the 1936 Flood Control Act, which gave the federal government authority to engage in public works projects "if the benefits to whomsoever exceed the cost." For our example we can set the ratio as follows:

Sum of benefits (gross or net) ÷ total cost

The ratios for our four investments are:

$10,000	*Total Benefits*		*Benefit/Cost Ratio*		
Investment	*Gross*	*Net*	*Gross*	*Net*	*Rank*
A	$10,000	$ —	1.	—	4
B	11,100	1,100	1.11	.11	3
C	11,524	1,524	1.1524	.1524	1
D	11,524	1,524	1.1524	.1524	1

Since the benefit/cost ratio is similar to an ROI, not too suprisingly the rank order is the same.

Discounted cash flow measures reflect the time value of money. Intuitively, we can say that money has a value that varies with time. Surely a dollar received today is worth more than a dollar received a year from now. It is worth more by the amount that could be earned by using that dollar between now and next year. Obviously, this amount depends on the interest rate used. Thus we have the two important factors in determining the time value of money: interest rate and time.

Discounted cash flow can best be understood by considering the lending operation in a bank. When money is borrowed, a fee is paid to the bank by the borrower for use of the money. This fee is expressed as an interest rate. The amount charged for the borrower's use of the bank's money over each compounding period is determined by multiplying the principal by the interest rate. This fee, called accrued interest, plus a portion of the principal is paid to the bank by the borrower at the end of each compounding period.

Assume that we borrow $1,000 at 5 percent interest with an agreement to pay off principal plus accrued interest over a four-year period in equal annual installments. Bank loans are commonly set up in this manner. The following table shows the lending situation:

Year	Principal at Start	Accrued Interest at 5%	Year-End Obligation	Annual Payment	Interest Paid	Principal Repaid
0	$ —	$—	$1,000	$ —	$—	$ —
1	1,000	50	1,050	282	50	232
2	768	38	806	282	38	244
3	524	26	550	282	26	256
4	268	14	282	282	14	268
						$1,000

The obligation is the sum of the principal plus accrued interest. We have simplified the example by rounding off to the nearest dollar; a bank would, of course, work out the numbers to the nearest penny.

The principles embodied in this type of loan are directly applicable to analysis of the desirability or profitability of a capital investment opportunity. That is, what is the maximum interest rate a project can afford in order to repay the money required for the investment when the earnings from the project are the sole source of income? Suppose a project costs $100,000, all spent essentially at one time; the project requires a year for construction after the money is spent and provides an annual cash income of $25,000 for the next five years. Putting this in table form, we get:

Year	Investment at Start	Interest at 5.823%	Earnings	Repayment on Investment	Investment at End
-1	$ —	$ —	$ —	$ —	$100,000
0	100,000	5,823	—	(5,823)	105,823
1	105,823	6,161	25,000	18,839	86,984
2	86,984	5,062	25,000	19,938	67,046
3	67.046	3,903	25,000	21,097	45,949
4	45,949	2,675	25,000	22,325	23,624
5	23,624	1,376	25,000	23,624	—
	$429,426	$25,000	$125,000	$100,000	

The interest rate is 5.8 percent, which results in 0 investment at the end of year 5. The total annual cash income of $125,000 repays the $100,000 original investment and provides a return of $25,000 on the investment; this $25,000 provides an interest rate of 5.8 percent on the money in the project. (The average investment over the six-year period is $429,426/6 = $71,571. At 5.8 percent, this average investment earns interest of $4,167 per year, or $25,000 for the six years.) This is the exact return an investor would realize on such an investment. Therefore, we have an economic indicator of the absolute economic worth of the investment proposal.

The DCF return can be determined by trial and error. It is simpler, however, to discount the earnings rather than compound the interest and calculate how much remains to pay back the principal. Hence the discounted cash flow concept. The principle behind the DCF calculation is that an investment involves the purchase of a series of future cash inflows. The DCF rate of return is determined by finding the interest rate by which each of the annual cash inflows must be discounted so that the discounted total equals the present value of the original investment. The computation is shown below. Parentheses have been used to indicate cash outflows.

Year	Year-End Cash Flow	Interest Discount Factor	Interest Net Present Value	Interest Discount Factor	Interest Net Present Value
-1	($100,000)	1.05	($105,000)	1.06	($106,000)
0	—	1.	—	1.	—
1	25,000	.9524	23,810	.9434	23,585
2	25,000	.907	22,675	.89	22,250
3	25,000	.8639	21,598	.8396	20,990

Year	Year-End Cash Flow	Interest Discount Factor	Interest Net Present Value	Interest Discount Factor	Interest Net Present Value
4	25,000	.8227	20,567	.7421	19,803
5	25,000	.7835	19,588	.7473	18,683
			$3,238		($689)

We can now calculate the net present values for our four investment alternatives:

$10,000 Investment	Present Value of Proceeds at 6%	Present Value of Outlay	Net Present Value	Rank
A	$ 9,430	$10,000	($570)	4
B	10,413	10,000	413	3
C	10,457	10,000	457	2
D	10,564	10,000	564	1

The DCF returns are as follows:

Investment	DCF Return (%)	Rank
A	—	4
B	10	1
C	9	3
D	10	1

Not surprisingly, the rank order for different DCF measures is not the same:

$10,000 Investment	Present Value of Proceeds at 30%	Present Value of Outlay	Net Present Value	Rank
A	$7,692	$10,000	($2,308)	3
B	8,343	10,000	(1,657)	1
C	7,437	10,000	(2,513)	4
D	7,842	10,000	(2,158)	2

Present value is the same as present worth. The net present value of a series of uneven outlays over time is influenced significantly by the assumed discount rate. This discussion has not taken into account the interest rate that should be used. Some would use cost of capital to the given company—but cost of capital is a difficult concept, as discussed previously.

Projects with the same original investment and DCF return do not yield the same present worth if the cash flow patterns are different.

TABLE 5-2. Financial criteria for decision making

Method	Definition	Computation	Advantages	Disadvantages
Payback period	Number of years until investment is recouped	If rate of flow is constant, payback equals investment ÷ net cash flow Otherwise, payback is determined by adding up the expected cash inflows until the total equals the initial investment	Simple to use and understand Makes allowances for risk attitudes Commonly known and used Useful as a constraint	Ignores cash flow beyond payback period Ignores timing within payback period Overemphasizes liquidity as an investment criterion
Accounting (or unadjusted) ROI	Ratio of average annual income after depreciation to average book value of investment	Average annual cash inflow minus average annual depreciation ÷ half initial investment.	Easy to compute and understand Commonly known and used	Ignores timing of cash flows

	Description	Formula	Advantages	Disadvantages
Net present value (NPV)	Difference between cash inflows and outflows discounted to the present at a given interest rate	$NPV = \sum_{t=1}^{T} \dfrac{F_t}{(1+i)^t}$ where F_t = net cash flow at time period t; i = discount rate; T = planning horizon	Takes time value of money into account; Easier to compute than ROI	Requires definition of a discount rate; Less intuitive than ROI
DCF (or internal rate of return)	Discount rate that makes the net present value of inflows and outflows equal to 0	The DCF return is determined by solving the equation $$\sum_{t=1}^{T} \dfrac{F_t}{(1+i)^t} = 0$$ where i is the rate of return on investment	Takes time value of money into account; Does not require definition of a cutoff rate; Intuitively appealing	Computationally complex (requires trial and error); Assumes other investment opportunities exist at same DCF; Does not consider size or scale of investment; Occasionally results in more than one discount rate, or none

This is because the original investment alone does not establish the amount of money on which the DCF returns are earned. The funds remaining in the project and earning the average DCF rate of return vary from year to year according to the cash flow pattern unique to the project.

Consider two projects with the same investment, same life, and same DCF return. In one project the cash receipts are uniform, in the other project the receipts are delayed by long startup and construction times and slow market development. The latter project would show a higher present worth because the compounding of interest on the original investment through the delay period results in higher average investment in the project. Of course, the risk would also be higher in the latter project.

Present worth is widely accepted as the best measure of the financial impact of a project on a company if all assumptions associated with the project are correct. Maximizing present worth with the available capital would yield the best financial result. However, most business situations involve substantial degrees of uncertainty. In this case, the DCF return is more useful by comparing it with an acceptable return criterion previously (and independently) developed, we can get a picture of the predicted margin of financial safety in the project, which can then be weighed against the risks involved. At equal present worths and equal risks, a moderate investment with a high DCF return would be much more attractive than a larger investment with a DCF return only slightly above the marginal value.

The examples showing the rank (relative attractiveness) of four investment alternatives may be confusing. Different methods of calculation, or different arbitrary factors, show first one alternative, then another, then a third or fourth, as being better than the others. How can that be? The fact is that the decision makers of a company, by long experience, develop a number of (personal) approaches for evaluating alternatives. They come to use the same combination of methods to measure each proposed investment. They learn with time how to weigh the different approaches and, perhaps, the different outcomes. Often their decision depends to a significant degree on the particular individual who is pushing for the investment alternative.

A comparison of the different techniques we have considered in this chapter is given in Table 5-2. All are useful in the right circumstances, provided one knows what one is doing. Since the discounted cash flow method of economic analysis is of such importance, it is examined in detail in Chapter 6.

chapter 6

The Discounted Cash Flow Method of Economic Analysis

Discounted cash flow analysis involves setting up a schedule of the amount and timing of all the cash received or spent during a project. Each cash flow is then multiplied by a discount factor, which accounts for the discount rate and the timing of the cash flow. Finally, the results are added up.

CASH FLOWS

"Cash flow" is an all-inclusive term embracing the economic elements that comprise profits as well as those comprising assets. Although the term does not embrace liabilities explicitly, it can be extended to cover some of them:

$$\text{Cash flow} = \sum \text{cash inflows} - \sum \text{cash outflows}$$

We have emphasized the basic difference between profits and assets in earlier chapters. The difference resides in the degree of liquidity or convertibility. Profits are, of course, completely liquid and may be converted into anything that money will buy, including assets. Assets, on the other hand, remain assets as long as the business exists. This is particularly true of fixed assets such as land, buildings, and equipment. As a matter of fact, when profits are converted into fixed assets, their value (except for land) almost literally evaporates through depreciation. Current assets (cash, inventories, and receivables) are interconvertible, but the combined amount of money in fixed assets cannot be used for any purpose other than borrowing unless the business shrinks or changes in character. The ROI and similar criteria are concerned with the irrevocability of money tied up in assets.

The concept of cash flow emphasizes the characteristic that profits and assets share: they are now, or once were, dollars. Generally,

Cash inflows = net income + noncash charges (depreciation, depletion)

Cash outflows = capital investment + working capital

Working capital, to most analysts, is the difference between current assets and current liabilities. By considering such monetary elements at the instant of receipt or conversion, while they are still "flowing" in the form of dollars, the DCF method gains two advantages over the ROI criterion:

○ *All monetary elements are additive.* The ROI criterion does not enjoy the "additive" advantage because it is concerned with the raw materials, buildings, and other assets that the dollars are converted into. Raw materials and buildings are like apples and oranges; they cannot be added. However, except for certain considerations of "financial image," it makes little difference what form the dollars are converted into as long as the legitimate long-term objectives of the company are met. The DCF method is designed to employ additive dollars.

○ *Actual timing of receipts and expenditures can be taken into account.* The ROI criterion also has the timing advantage, to a limited degree, by virtue of its ability to specify the period over which the profits are being considered. However, the ROI criterion cannot account for a difference in the timing of an expenditure for a fixed asset, which may extend over several years. Nor can it adequately account for different profiles of profit over several years. The choice between alternative plans of action may well depend on these differences, which the DCF method is specifically designed to handle.

_____ FUNDAMENTALS OF DISCOUNTING

In order to predict the economic effect of a proposed course of action, we must recognize effects that will occur over many years in the future. If a proposed course of action involves the construction of a new plant for the manufacture of a salable product, the effects will not be the same every year.

In the year the action is taken, there will be an expenditure for buildings and equipment but little or no sales. In subsequent years, sales receipts will increase, because a plant is rarely utilized to capacity when it goes on stream. Likewise, expenditures for raw materials, utilities, and possibly labor will increase as more product is made. The ROI for the first year will obviously be very low, if not negative. It will probably increase for several years until capacity is reached, but it may then decrease if competitors force a reduction in the sales price of the product or if prices cannot keep pace with escalating costs.

What would be the meaning of a series of ROIs, one for every future year that was predictable? Although such a series might show that a criterion of, say, 20 percent was ultimately achieved, it would not tell us *when* the criterion was achieved (or for how long). We would prefer that it was achieved sooner instead of later. Intuitively, we reason that the sooner a profit is obtained, the earlier it can be put to use to expand the company or to be paid out to stockholders, who could then put it to new productive uses. This argument recognizes that, literally, time is money. Or to put it in financial language, money has a time value. But how do we quantify the advantage of obtaining a profit sooner? This is where the concepts of discounting and present value enter the picture.

Discounting is the exact opposite of compounding. In compounding, if $1 is invested now at 10 percent interest, it will be worth $1.10 a year from now and $1.21 two years from now. But this dollar cannot be invested until it is received, and every year the receipt is delayed interest is lost. If the receipt of $1 is delayed for one year, the $.10 interest is lost. The dollar is worth only $1 instead of the $1.10 it would have been worth if it had been received a year sooner. A dollar received one year late is therefore worth only

$\$^1/_{1.10}$ = 90.9 percent as much as a dollar received a year sooner. Similarly, a dollar that is two years late is worth only $\$^1/_{1.21}$ = 82.6 percent as much as a dollar received two years sooner. The decimal equivalents of these percentages, .909 and .826, are called *discount factors.*

The discount factor depends on the interest rate at which a given sum could have been invested if it had been received sooner. For example, the discount factor applicable to a dollar delayed for one year at 6 percent interest is $\$^1/_{1.06}$ = .943. At lower interest rates the discount factors would be higher, since less interest is lost. If the interest rate were 0, the discount factors for all years would be 1.

Mathematical formulas can be derived to show these present and future values at any interest rate and for any period of time. However, the numbers have already been worked out in the form of financial tables or compound-interest tables, which are available in many financial and engineering economics texts. A complete set of tables would include:

Single-payment compound-amount factors, or the compound amount of $1. This is the amount $1 will grow to in a given number of years at a given interest rate.

Single-payment present worth factors, or the present worth of $1 received in the future. The present worth is lower than $1 because a lower amount could accumulate compound interest during the interim period to yield $1 at a future time. These are the reciprocals of the single-payment compound-amount factors.

Uniform annual series compound-amount factors, or the compound amount of $1 per year added to the accumulation at compound interest.

Sinking fund deposit factors, or the annual payment that accumulates to $1 at the end of the period. These are the reciprocals of the uniform annual series compound-amount factors.

Capital recovery factors, or the annual payment that $1 will buy.

Uniform annual series present worth factors, or the present worth of $1 per year. These are reciprocals of the capital recovery factors.

Different tables may be based on different time intervals. For example, the discount factors at an 18 percent rate are:

Year	Continuous	Year End	Midyear
0	1.	1.	1.
1	.8353	.8475	.922
2	.6977	.7182	.781
3	.5828	.606	.622
4	.4868	.5158	.56

We have shown the extremes or maximum difference; if we discounted quarterly or monthly, the numbers would lie between the two numbers shown in the continuous and midyear columns.

Generally, no attempt is made to predict the date of a cash flow other than to name the calendar year in which it occurs. For this reason, monthly or weekly discount factors are not required. Even when it is known that a cash flow such as sales will occur every day throughout a year, the total amount of the cash flow in that year is generally discounted by the single discount factor applicable to that year. This practice tacitly assumes that the yearly flow of cash occurs at the beginning of each year. It also assumes that money is compounded annually. This assumption is not universally true, inasmuch as semiannual and continuous compounding are also employed. However, annual compounding (and discounting) results in great arithmetic simplification. Any inaccuracies created by these simplifying assumptions tend to cancel out, since the same methods are used for each investment evaluation. (Exceptions may be made for specific cases.)

The significance of discounting is that if the discounted value of a delayed dollar is received today, it will be worth just as much as the whole dollar will be worth when it is actually received. A dollar received one year late has a discount factor of .909 at 10 percent interest, since $.909 invested now at 10 percent interest will be worth $1 a year from now. A discount factor, therefore, permits the calculation of the equivalent value of a dollar received at any future time. By stating the equivalent value of all future dollars in relation to a single basing point (the present), we can sum the equivalent values of many different dollars received at many different times and thereby determine the total real effect of those dollars.

Let us see how discounting is actually done. Assume first that an investment of $1,000 will be made in 1980 to construct a new plant:

	1980	1981	1982	1983
Investment	$1,000	—	—	—

In 1980 and in the following three years, net sales are assumed to be:

	1980	1981	1982	1983
Net sales	$100	$700	$900	$1,100

The sum of all the costs and taxes in each year is assumed to be:

	1980	1981	1982	1983
Costs and taxes	$200	$400	$500	$600

Further, assuming that the time value of money is approximated by a 10 percent interest rate, the discount factors applicable to the four years in question are:

	1980	1981	1982	1983
10% discount factors	1	.909	.826	.751

Conventional nomenclature identifies year 0 as the year to which a discount factor of 1 applies. Year 0 can be the year when the decision is made to embark on a course of action or the year when the first commitment of funds is made. In the example under discussion, 1980 is assumed to be year 0.

If we multiply each cash flow by the discount factor that applies to the year in which the cash flow takes place, we obtain the present value of each cash flow. The algebraic sum of the present values of all cash flows is the net present value (NPV).

Let us consider the possibility of combining some of the cash flows to reduce the number of multiplications. Cash flows in different years cannot be combined, because they must be multiplied by different discount factors. Cash flows in the same year can be combined, provided we recognize which of the cash flows are receipts of money and which are expenditures.

If a $1 receipt and a $1 expenditure took place in the same year, a company would be neither richer nor poorer. If one of the expenditures had a negative sign, the result would be what we expect: 0. Therefore, by convention, minus signs or parentheses are used to indicate expenditures of money. The algebraic sum of the cash inflows and outflows in each year is the annual cash flow in that year.

Restating the cash flows with the proper signs, and omitting the dollar signs as is customary, we can calculate the NPV as shown in Table 6-1. The NPV is −$121. This means that, if money has a time value of 10 percent, the course of action will not be attractive if it stands alone.

TABLE 6-1.   Net present value

	1980	1981	1982	1983
Investment	(1,000)	—	—	—
Net sales	100	700	900	1,100
Costs and taxes	(200)	(400)	(500)	(600)
Annual cash flow	(1,100)	300	400	500
10% discount factors	1	.909	.826	.751
Annual present value	(1,100)	273	330	376
Cumulative sum of annual present values	(1,100)	(827)	(497)	(121)
				NPV

This conclusion should lead to a closer examination of cash flows. What can be done to make the course of action attractive? What alternatives exist? This sort of reexamination forms the heart of a good analysis. It can point the way to a profitable opportunity that would have been lost if the figures were accepted as unalterable.

In the above example, we could raise the NPV by reducing the investment. Observe that the discount factor for the year 1980 is 1. For every dollar that the investment can be reduced, the NPV will increase by a dollar. A reduction of only $121 would suffice to raise the NPV to 0, which may be sufficient *if* there is no risk attached to the anticipated income from net sales. Perhaps the plant could be redesigned to eliminate components that are not absolutely necessary. Perhaps less expensive components could be used. Perhaps the investment estimate itself is incorrect.

Alternatively, we could raise the NPV by increasing the income from net sales. Assuming it is not possible to raise the selling price, we would have to increase volume, which would also increase costs. However, costs do not generally increase in proportion to volume. For example, if net sales were increased from $700 to $900 in 1981, costs might increase from $400 to only $500. This would add $100 of cash flow in 1981. Discounted with a factor of .909, this extra $100 would contribute an additional $91 to the NPV. If the same increases occurred in 1982, the extra $100 cash flow in that year would contribute $83 to the NPV. What is the likelihood that the original estimates of selling price and sales volume were correct? Perhaps this opportunity is better than it appeared at first. A caution is in order here. Any number used in an analysis must reflect a plan or some market information. Wishful thinking is not enough.

We can also increase the NPV by decreasing costs. Perhaps the prices assumed for raw materials in the original estimate were too high; perhaps materials can be used more efficiently. Perhaps fewer people can operate the plant that was assumed. As with an increase in net sales, a decrease of $100 in costs in 1981 would contribute $91 to the NPV, and so on. The critical question is: What is the likelihood that the original estimates of costs were correct?

Another way to increase the NPV is to sell the product for one or more years after 1983. Suppose, for example, that 1984 looked just like 1983:

Investment	—
Net sales	1,100
Costs and taxes	(600)
Annual cash flow	500
10% discount factor	683
Annual present value	342

Clearly, an extension of the "life" of a profitable course of action is one way to increase the NPV. However, beyond the question of whether a market for the product will indeed exist, this procedure raises a fundamental issue: What life should be assumed in an analysis when there is no known limitation? Almost any borderline course of action can be made to look profitable if we extend the number of years embraced by the analysis. Similarly, courses of action that are good can be made to look superlative.

Obviously, the farther into the future the prediction is extended, the less confidence can be placed in the prediction. If there is reason to believe that the profitable life of a course of action will in fact be limitless, there is nothing wrong with extending the analysis to any number of years, *provided* this is brought unmistakably to the attention of the decision maker. In general (there are exceptions), a course of action requiring more than 15 years of cash flows to generate a positive NPV is of doubtful worth. Moreover, any attempt to "force" an NPV by a gross extension of the analysis is open to suspicion. On the other hand, we want to take into account as much of the future as we can attribute valid meaning to. To justify investments in long-term assets, we must be able to make reasonably confident predictions of the next 5 or 10 years. A "standard life," if there is such a thing, is commonly accepted to be 15 years for manufacturing facilities.

Finally, we can increase the NPV by using discount factors for a lower rate of interest. That is, the estimated time value of money may not be as high as 10 percent. If we use an interest rate of 4 percent, the annual cash flow of $300 in 1981 would be discounted by a factor of .962 instead of the factor of .909 associated with the 10 percent interest rate. An increase in the NPV of $300 × (.962 − .909) = $16 would result for the year 1981 alone. A revised analysis for the entire course of action, using discount factors for a 4 percent interest rate, is given in Table 6-2.

Changing the discount rate has two important implications. The first is that the time value of money is subject to variation. (This implication is discussed later in the chapter.) Thus a "standard" interest rate must be established for the sake of placing all NPVs on a consistent basis. This standard interest rate has ranged from 10 to 20 percent (or several times the cost of capital) for U.S. industry. It is called the *marginal rate of return,* or cost of capital, and is abbreviated MRR. Strictly, there is no such thing as a "4 percent NPV." By definition, an NPV is calculated at the marginal rate of return criterion selected by a company. However, the calculation of the sum of the annual present values (the cumulative present values) for a number of different interest rates is a useful analytical tool, as we will see below.

TABLE 6-2. Net present value with 4 percent discount factors

	1980	1981	1982	1983
Investment	(1,000)	—	—	—
Net sales	100	700	900	1,100
Costs and taxes	(200)	(400)	(500)	(600)
Annual cash flow	(1,100)	300	400	500
4% discount factors	1	.962	.925	.889
Annual present value	(1,100)	289	370	445
Cumulative sum of annual present values	(1,100)	(811)	(441)	4
				NPV

The second implication of changes in the interest (discount) rate for most analyses is that there is some interest rate that will produce a cumulative present value of 0 for a particular course of action. That interest is, of course, the *DCF rate of return* for that course of action. In the foregoing example, note that the DCF return is very close to 4 percent, since a change of 1 percentage point in either direction would produce a cumulative present value farther away from 0. (Check this for your own satisfaction.)

The calculation of a DCF return is a trial-and-error process. It proceeds by assuming different interest rates and calculating the corresponding cumulative present values. This calculation may be made less tedious by plotting cumulative present value versus interest rate on a graph. In the preceding example, three points for the curve are already available:

Interest Rate (%)	Cumulative Present Value
10	(121)
4	4
0	100

The curve is plotted in Figure 6-1. At an interest rate of 0 percent, the discount factor for every year is 1. Therefore, the cumulative

Figure 6-1. Cumulative present value vs. interest rate

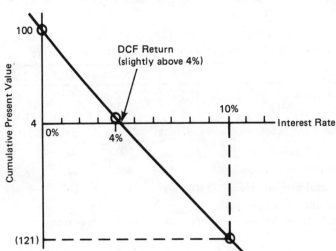

present value is simply the algebraic sum of the undiscounted annual cash flows. Because the formula for calculating discount factors is similar to a hyperbolic function, the curve of cumulative present value versus interest rate is not really a straight line. (An exception is a course of action with only one annual cash flow, occurring in year 0. In that case, the curve is a horizontal straight line and there is no DCF return.)

The significance of a DCF return has been argued in the literature to an unparalleled extent. It means many things to many people, and the key issue seems to be the nature of "interest." Some say that interest is literally money paid for the privilege of borrowing money. Others say that interest must also include dividends paid to the stockholders of a corporation. Still others say that interest has no literal connotation when used for the purpose of discounting, but is only a means of stating numerical criteria.

For the moment we can use a popular definition of the DCF return: "The DCF return is the maximum rate of interest that a company *could* pay to borrow the funds necessary to finance a course of action. By the time the course of action is complete, all the borrowed funds will have been paid back and the company will have broken even." This definition, while generally correct, fails when the course of action has no DCF return, a negative DCF return, or multiple DCF returns (which can result from certain cash flow patterns as discussed later in the chapter).

Often it is best to consider the calculated interest rate (DCF return) on a contemplated investment as merely a numerical criterion. It has no real-life meaning. For example, a projected 15 percent DCF return means that the *project* is "borrowing" money from the parent company at an interest rate of 15 percent. The project pays 15 percent per year on the unrecovered balance until the debt reaches 0 at the end of the project life. But the parent company does not really get the benefit of the 15 percent interest payment—unless it can reinvest the proceeds "paid back" by the project in other enterprises that actually return 15 percent year by year.

_____ PRESENT VALUE OF MAXIMUM CASH COMMITMENT

Like any other measure of business performance, an NPV must be qualified according to how it was derived and how vulnerable it is to changes in the assumptions regarding investment, net sales, and so on. Two measures of performance that help to characterize the NPV are the *present value of maximum cash commitment* (MCC) and the *payback.*

For example, note that in Table 6-3 the cumulative present value decreases to a maximum negative value of $1,409. This is the MCC. From this point, the cumulative present value increases until it changes sign in 1986. The year 1986 is the payback. In a manner of speaking, the MCC is the depth of the financial hole that a company is put into by a selected course of action. The payback is the year the company climbs out of this hole. Recall that in Table 6-1 the NPV was negative. If the company had undertaken the project, it would never have climbed out of that hole and thus would have had no payback. There was, however, an MCC of $1,100.

The NPV and MCC are related in much the same way that the ROI income and total utilized investment (TUI) are related. Although it is not customary to calculate the ratio of the NPV and the MCC, the MCC provides a measure of the resources that will have to be committed to achieve the profit measured by the NPV.

Payback has no counterpart in ROI terminology. "Payout" has sometimes been used. It is the "payback" that would be derived from cumulating the undiscounted annual cash flows. Payback indicates what might be called the risk of exposure. The longer it takes to reach payback, the longer a company is exposed to a net loss on a course of action. At the instant of payback the NPV is 0.

TABLE 6-3. Cumulative present value

	1983	1984	1985	1986	1987
Investment	(500)	(1,000)	—	—	—
Net sales	—	—	2,000	2,000	2,000
Costs and taxes	—	—	(1,000)	(1,000)	(1,000)
Annual cash flow	(500)	(1,000)	1,000	1,000	1,000
10% discount factors	1	.909	.826	.751	.683
Annual present value	(500)	(909)	826	751	683
Cumulative present value	(500)	(1,409)	(583)	168	851
					NPV

The interplay among NPV, MCC, and payback can best be seen by plotting the cumulative present values into a curve. The curve for the data in Table 6-3 is shown in Figure 6-2. The curve is not a typical one. In most cases, things do not get off to such a fast start, and the curve is more rounded (see Figure 6-3*a*).

Several considerations enter into decisions that utilize NPV criteria. A study of three typical cases will clarify these issues. First, suppose that two mutually exclusive opportunities present themselves. Both have equal risks surrounding income and cost. These opportunities differ in only one way. Opportunity A has a greater NPV than Opportunity B (see Figure 6-3*b*). The choice is clear. If either A or B must be chosen, A is preferred.

Next, suppose A and B differ only in the fact that B has a greater MCC than A (see Figure 6-3*c*). Opportunity A wins on two counts. It requires a smaller commitment of resources to achieve the same NPV. And the NPV is the same only because the project life is such that both curves coincide. If the projects were extended one more year (the dashed lines in Figure 6-3*c*), A would have a higher NPV. *Opportunity A has more potential.*

Figure 6-2.   NPV, MCC, and payback

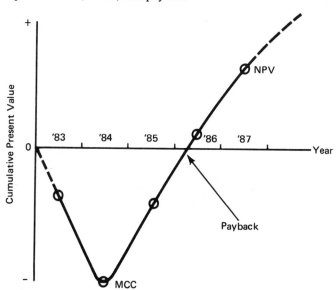

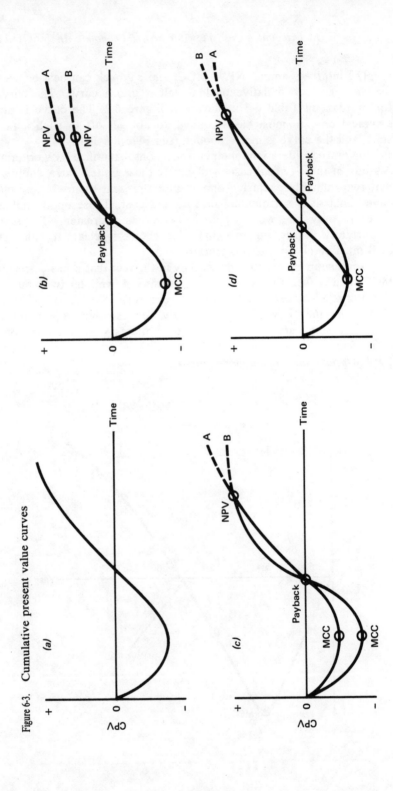

Figure 6·3. Cumulative present value curves

Finally, suppose A and B differ only in the fact that B has a later payback than A (see Figure 6-3*d*). Here the decision is not obvious. Opportunity A has less risk of exposure because of its earlier payback. But Opportunity B has greater potential.

The point of this discussion is that, even in simple cases where all other things are equal, decisions can be difficult. Imagine, then, how difficult decisions are in ordinary cases where things are not equal. The conclusion to be drawn here is that there is no substitute for a complete display of the facts.

_____ MOVEMENT OF YEAR 0

Year 0 is defined as the year in which a decision is made to embark on the action being analyzed, usually with a commitment or expenditure of funds. In most cases, 0 is the current calendar year, because an analysis is made to be acted upon. However, when an analysis is being made to explore a possible opportunity several years hence, year 0 is the date the future decision will be made. By the same token, when the consequences of decisions made several years in the past are reviewed, year 0 is the year the decision was actually made.

The reason for equating year 0 with the year of decision is that the cash flows usually begin with that year. If several years will pass between the time of analysis and the time of decision, and the analysis year is treated as year 0, the first cash flow will be discounted by a relatively low factor. Subsequent cash flows will be discounted by even lower factors. For example, assume that the year of analysis is 1975 and that the analysis is concerned with constructing a plant to produce a new product for which funds will not be committed until 1980. If year 0 is placed in 1975, the first cash flow in 1980 will be discounted with a factor of .621. The final cash flow, say 14 years later, in 1994, will be discounted by a factor of .164. However, if year 0 is placed in 1980, the first cash flow will be discounted with a factor of 1 and the final cash flow will be discounted with a factor of .263 (see Table 6-4).

The NPV calculated with year 0 in 1975 is not comparable to the NPVs of other analyses made in 1975 for immediate action. The frame of reference is lost. The calculated NPV does not, in effect, transport the decision maker ahead to 1980 so that he can say: "This is the NPV I will see when I am really faced with the decision.

TABLE 6-4. Change of year 0 in discount cash flow analysis

Year 0 in 1975			Year 0 in 1980	
Annual Cash Flow	10% Discount Factors	Calendar Year	10% Discount Factors	Annual Cash Flow
	1	1975		
	.909	1976		
	.826	1977		
	.751	1978		
	.683	1979		
	.621	1980	1	
	.564	1981	.909	
	.513	1982	.826	
	.467	1983	.751	
	.424	1984	.683	
	.386	1985	.621	
	.35	1986	.564	
	.319	1987	.513	
	.29	1988	.467	
	.263	1989	.424	
	.239	1990	.386	
	.218	1991	.35	
	.198	1992	.319	
	.18	1993	.29	
	.164	1994	.263	

This NPV is of the same order of magnitude as others upon which I am basing decisions today. Therefore, I can look forward with confidence to what my future decision will be.''

But suppose that, after calculating the NPV with year 0 in 1980, the decision maker realizes that there will be a substantial profile of developmental costs between 1975 and 1980. He suspects that these costs may wipe out a large portion of the NPV of this opportunity. Now he must calculate the NPV of the developmental costs with year 0 in 1975, because these costs must be decided on immediately. He cannot subtract the NPV of the developmental costs from the NPV of plant production since each employs a different year 0.

One solution to this problem is to recalculate the NPV of the cash flows that will occur in 1980 and following years using 1975 as year 0. This procedure can be laborious. The same results can be obtained by multiplying the 1980-base NPV by the ratio of the

1980 discount factor to the 1975 discount factor $^{.621}/_1$ with 1975 as year 0.

For any given interest rate, each successive discount factor is the same multiple of the factor that precedes it. The multiple is the year 1 discount factor. This follows from the formula for calculating discount factors, which is:

$$\frac{1}{(1 + i)^n}$$

where i = the decimal interest rate
n = the years of time delay

The discount factor for year 1 is:

$$\frac{1}{(1 + i)^1} = \frac{1}{(1 + i)}$$

The discount factor for year 2 is:

$$\frac{1}{(1 + i)^2} = \frac{1}{(1 + i)} \times \frac{1}{(1 + i)}$$

year 1 factor   year 1 factor

The discount factor for year 3 is:

$$\frac{1}{(1 + i)^3} = \frac{1}{(1 + i)} \times \frac{1}{(1 + i)^2}$$

year 1 factor   year 2 factor

The discount factor for year 4 is:

$$\frac{1}{(1 + i)^4} = \frac{1}{(1 + i)} \times \frac{1}{(1 + i)^3}$$

year 1 factor   year 3 factor

Thus each successive factor will be equal to the preceding factor times the year 1 factor.

If a series of cash flows is multiplied by a certain train of discount factors to obtain an NPV, moving the train of factors backward by one year will decrease every present value in the series by the same multiple: the year 1 factor. The NPV will, of course, decrease by the same factor. If the new discount train is moved backward an additional year, the previously decreased NPV will be decreased again by the same multiple: the year 1 factor. This two-year movement has resulted in the year 1 factor being applied twice—that is, squaring the year 1 factor. This is exactly equivalent to applying the year 2 factor to the original NPV.

Similarly, if a train of discount factors is moved forward by one year, every present value in the series will increase by the same multiple: the reciprocal of the year 1 factor. If the discount train is moved forward by two years, every present value in the original series will be increased by the square of the reciprocal of the year 1 factor. This is exactly equivalent to the reciprocal of the year 2 factor. Out of this come two rules:

Rule 1: $\dfrac{\text{NPV with time 0 in year } x}{\text{NPV with time 0 in year } y} = \dfrac{\text{discount factor for year } y}{\text{discount factor for year } x}$

Rule 2: Regardless of the movement of year 0, the NPV does not change sign.

These rules are also applicable to years preceding the current year. Often after an action has been in motion for several years, some revision is proposed. An NPV for the revised plan is calculated, using the current year as year 0 and considering only those cash flows from the current year forward. The question is then asked: What does the entire course of action look like, including past years? An NPV for the past years can be calculated, with the original year of decision as year 0. But this cannot be added to the NPV calculated for future years with the current year as year 0. The original year of decision must be treated as year 0, and the current-year NPV must be adjusted backward to that base. Obviously, the current-year NPV will be reduced.

Sometimes the question is asked: What does the entire course of action look like, including past and future years, with the *current year* as year 0? In this case, the answer is meaningless. An NPV serves no decision-making purpose when it is based on year 0 that occurs in the middle of a series of cash flows. Still, a numerical answer can be given if necessary, employing the *discount factors for past years*. The factors for past years are not technically "discount" factors. They are compound factors and represent the denominators of the corresponding discount factors $(1 + i)^n$. Thus they are the reciprocals of the corresponding discount factors. The compound factors at 10 percent interest are:

Past Years					Current Year	Future Years				
5	*4*	*3*	*2*	*1*	*0*	*1*	*2*	*3*	*4*	*5*
1.61	1.464	1.332	1.211	1.1	1	.909	.826	.751	.683	.621

DIFFERENTIAL CONVENTIONS

In the decades since DCF has become a generally accepted tool of financial analysis, anomalous or unusual DCF returns have been calculated. For example, several values of DCF may be produced from one set of cash flow data. Sometimes these appear to contradict the basic theory of the time value of money, which the DCF method rests. However, all have been satisfactorily explained, and in the process some fundamental characteristics of cash flow patterns have been recognized. We need to be aware of cash flow patterns that can lead to seemingly anomalous DCF returns.

The basic cause of anomalous or unusual DCF returns is the differential nature of economic analyses. A differential implies two alternatives, both of which may involve cash flows. In order to keep proper account of the cash flows on both sides of a differential, the analyst must pay careful attention to algebraic signs. DCF sign convention considers receipts of money as positive and expenditures of money as negative; it is also customary to refer to positive flows as "favorable" and to negative flows as "unfavorable."

In a differential analysis, it is important to view the differential from the standpoint of one alternative. That is, the cash flows of both alternatives must be viewed as favorable or unfavorable with respect to the alternative whose standpoint is taken. An unfavorable cash flow, such as the standpoint alternative's own investment, is given a negative sign. A favorable cash flow, such as the *other* alternative's investment, is given a positive sign. Therefore, when the cash flows of both alternatives are compared in the differential (one alternative versus or "minus" another), all signs in one of the alternatives must be reversed.

A differential should be named in a manner that makes clear the identity of the two alternatives. If it is generally understood that there are absolutely no cash flows in one alternative (for example, a proposed plant to make a new product), the concise name "new plant" is preferable to "new plant versus nothing." However, when both alternatives are real, a more expressive name such as "new plant X versus new Plant Y" is indicated. The following differentials require two real alternatives:

Make versus buy
Make instead of buy
Depart from buy to make
Make cash flows minus buy cash flows

The last expression clearly indicates the need to reverse the signs in the "versus" alternative.

The usual standpoint alternative is the one having the larger initial expenditures. Generally, it is listed first in the name of the differential so that the differential will exhibit a normal pattern of annual cash flows. In a normal pattern, negative cash flows in early years are followed by positive cash flows in later years:

$$- - + + +$$

The classic example of a normal pattern is a "new product versus nothing" differential, which involves expenditures that are justified by subsequent gains. Cost reduction differentials are also of the normal type.

The trend of the cumulative present value in a normal differential is downward to the right as the interest rate increases (see Figure 6-4). At 0 interest rate, where the discount factor for every year is 1, the cumulative present value is the algebraic sum of the undiscounted annual cash flows. At higher interest rates, the discount factors for early years remain close to 1, but the discount factors for later years become progressively smaller. Higher interest rates reduce the positive cash flows more than they reduce the negative cash flows. As a consequence, the cumulative present value tends to become smaller, and may even become negative, as the interest rate increases.

An example of an abnormal differential is the construction of a plant by a competitor to make a product that will yield additional income. From our point of view, the competitor's expenditure for a plant is favorable, but the later additional income is unfavorable. For us, the relative position of annual cash flows is:

$$+ + - - -$$

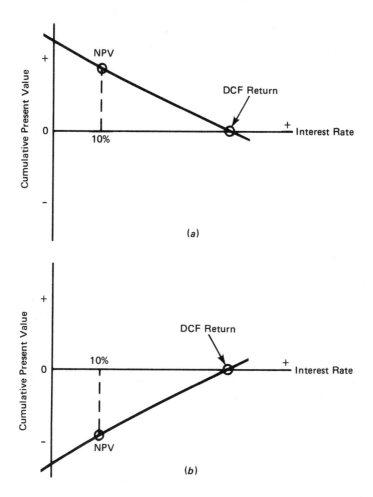

Figure 6-4. Cumulative present value. (a) Normal differential. (b) Abnormal differential

If the venture is profitable to the competitor, it may, for that reason, be unprofitable to us. *Our* cumulative present value at the marginal rate of return will, therefore, be negative (see Figure 6-4). In order for the positive cash flows to achieve ascendancy over the negative cash flows, higher interest rates must be used.

An analysis of the changes to a company's economic picture from an improvement in a competitor's economics may seem academic.

But whenever we recognize two alternatives in any opportunity, we automatically place them in competition. The direction of the cumulative present value trend depends on the alternative from which we take our point of view. Unless the point of view is clearly defined, the result may be misinterpreted.

Consider two other alternatives, both involving the same construction project designed to produce additional income. The basic difference between the alternatives is timing: one plant is built earlier and produces additional income earlier. From the point of view of the earlier plant, the earlier expenditure is unfavorable. However, subsequent annual cash flow differences will be favorable to the earlier plant because it will be in profitable operation while the later plant is being constructed. The cumulative present value trend is therefore illustrated by the normal differential. From the point of view of the later plant, the expenditures for the earlier plant are favorable, but the subsequent cash flow differences are unfavorable. The cumulative present value trend is therefore illustrated by an abnormal differential. The earlier plant's point of view is normal in that we are trying to justify a present expenditure by a future gain. The later plant's point of view is abnormal in that we are trying to justify delaying a gain by a present saving.

In practice, many alternatives are hybrids of the types discussed above. For example, a comparison may be made between the early construction of a small, less profitable facility and the later construction of a large, more profitable facility. With each alternative thus having both normal and abnormal characteristics, it is difficult to predict which cumulative present value trend will be observed from each point of view. A quantitative criterion to guide the selection of point of view would be highly desirable, but none has been found to be infallible. Therefore, the recommended procedure in a differential DCF analysis is to plot enough points to define unmistakably the direction of the cumulative present value trend.

Although the cumulative present value curves we have shown occupy the right-hand quadrants of the graphs, they can be located in any quadrant. In Figure 6-5a, for example, the direction of the cumulative present value trend is the same as that for a normal differential, indicating that the normal standpoint was taken. However, the negative cumulative present value at a 10 percent interest rate means that the standpoint alternative is less profitable than the other alternative. In this case, the standpoint can be reversed by reversing the signs of all cash flows. The cumulative present value curve will

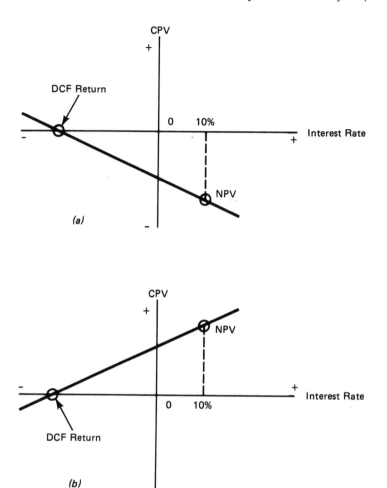

Figure 6-5.   Differentials showing negative DCF returns

then pivot on the DCF return point, as shown in Figure 6-5*b*.

Negative DCF returns require an explanation. Recall that an interest rate is universally regarded as a charge paid by a borrower to a lender. When the interest rate increases, borrowing money becomes more expensive, and the borrower's economics are adversely affected. We can say that a trend sloping downward to the right exhibits the adverse effect of an increasing interest rate on the *borrower's* economics. If the interest rate were 0, there would be no charge for the loan. The borrower would have to repay only its face value. If the interest rate were negative, the lender would pay the borrower to use the money. The borrower would then have to repay less than the face value of the loan.

Reversing the signs of all cash flows (that is, changing the standpoint of the differential) produces a curve sloping upward to the right. We can say that such a trend exhibits the beneficial effect of an increasing interest rate on the lender's economics. Reversing the signs of the cash flows has no effect on the sign of the interest rate. A negative interest rate therefore improves the borrower's economics and acts adversely on the lender's economics. To put it another way, whether a negative DCF return is good or bad depends on the alternative taken as the standpoint.

When two *real* alternatives are being compared in a differential, the interpretation may not be so obvious. The trend line of the differential may be where we would least expect it. *The DCF return for the differential is always the interest rate at which the trend lines for two real alternatives cross. If they do not cross, the differential does not have a finite DCF return.* DCF analysis should not be burdened with distinctions between "normal" and "abnormal" or between "borrower" and "lender." The meaning of the numerical results is usually quite clear if each of the alternatives is analyzed separately.

Differential trends that are horizontal straight lines are produced when both alternatives have equal annual cash flows in every year except year 0. The differential in that case has only one annual cash flow. This could easily occur if there were two ways of building the same plant and one was less expensive. Although a rather special case of "no return," this is the basic reason for the absence of a DCF return: *if every annual cash flow in a differential is of the same sign, there will be no DCF return.* The trend lines will appear as shown in Figure 6-6.

We can conclude that the existence of a DCF return depends on part of the annual cash flows in a differential being of one sign and part being of the opposite sign. Further, recall that in our examples there has been only one DCF return and only one *change* in sign:

$$-\ -\ +\ +\ +\ \qquad \text{or} \qquad +\ +\ -\ -\ -$$

From this it may be correctly inferred that *there are as many DCF returns in a differential as there are changes in the sign of the annual cash flow.*

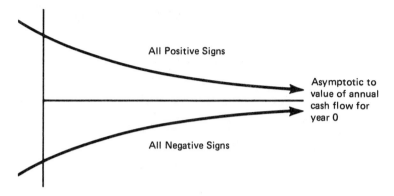

Figure 6-6.   Differential showing no DCF return

Suppose we are considering two competing investment projects. Their net cash flows (in thousands of dollars) are:

Year	Project A	Project B	Project A Minus Project B
0	(1,000)	(750)	(250)
1	525	465	60
2	425	345	80
3	325	225	100
4	225	105	120
DCF return	19.7%	25.1%	14.6%

Project A has a DCF return of 19.7 percent; Project B, 25.1 percent. If the minimum acceptable return were 10 percent, both projects would be acceptable. However, the problem often is to select the *better* project, since it may not be possible to participate in both projects. It is not sufficient simply to select the project with the higher return. The pertinent question is whether the additional investment required for Project A produces enough additional cash inflow to be attractive, because it is advantageous to commit additional funds only if they can earn a return higher than the minimum acceptable standard. The appropriate comparison is then between the rate of return on the incremental investment and the minimum acceptable return.

Figure 6-7 shows the relationship between the various DCF returns for the two competing projects. If 10 percent were the applicable standard, Project A would be more attractive because of the 14.6 percent return on the incremental investment. If, however, the mini-

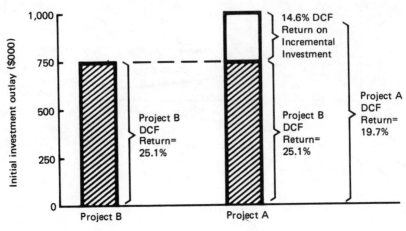

Figure 6-7.   DCF returns: Project A vs. Project B

mum standard were above 14.6 percent, Project B would be more attractive.

_____ PRESENT VALUE PROFILE ANALYSIS

If more than two competing alternatives are involved, the calculation of an incremental DCF return for each pair of alternatives can be cumbersome. In such a case, or in cases where a variety of different assumptions for each alternative must be tested, the present value profile method is a helpful tool.

Mutually exclusive projects can be compared by calculating the present value amounts for each alternative. These amounts are obtained by discounting each cash flow at a single discount rate corresponding to a minimum acceptable return appropriate for the investment opportunity. Considering only the economics of the proposals, the alternative showing the *highest positive* present value at the minimum acceptable standard for such an investment would be preferred. However, because a minimum acceptable rate of return is a matter of judgment involving the various elements of risk in a particular proposal, a graphic portrayal over a range of discount rates is preferable.

A profile of present value dollars versus discount rates is prepared by discounting the net cash flow of each alternative at various interest rates and plotting the results on a graph. Figure 6-8 shows a present

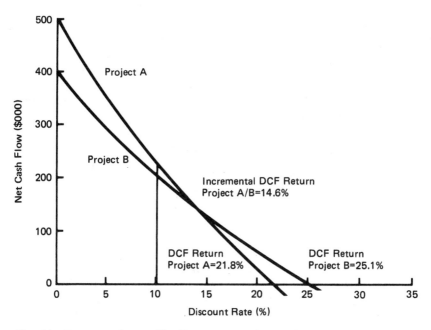

Figure 6-8.   Present value profile: Project A vs. Project B

value profile constructed from the net cash flows of competing Projects A and B, discussed previously. For simplicity only two alternatives are shown; in practice all competing alternatives could be plotted on the graph.

Recall that the DCF return is the rate that discounts the net cash flow to 0 present value. The discount rate at which the curve crosses the 0 present line is therefore the DCF return—in this case, 21.8 percent for Project A and 25.1 percent for Project B. These are the estimated DCF returns should Project A or B be implemented, and in each case the return is relative to the alternative of not committing any funds at all.

The incremental DCF return for Project A relative to Project B is the intersection of the two present value curves, or 14.6 percent (see Figure 6-7). The amount of present value advantage in dollars can be determined from Figure 6-8 for each alternative over a range of discount rates. At 10 percent, for example, the present values of Projects A and B are $226,000 and $199,000 respectively, indicating that Project A is more attractive by $27,000.

Note that if there is a positive present value for *each* alternative at the minimum acceptable return level, either project is acceptable. (In such a case, the DCF return for each project will be higher than the minimum acceptable standard.) However, when it is necessary to choose between two competing, risk-equivalent alternatives, the investor selects the alternative with the *highest positive* present value at the minimum acceptable return level.

A present value profile is a useful tool for comparing and ranking mutually exclusive projects, for the following reasons:

○ All the alternatives under consideration may be compared using the same assumed minimum return standard. It is possible to test how different judgments about the minimum acceptable rate of return can influence the choice among competing alternatives.

○ A present value profile displays the financial relationships among the alternatives, including all the pertinent DCF returns, and highlights their relationship to an assumed minimum acceptable return standard.

○ A profile indicates clearly situations in which even though neither alternative is attractive, the incremental investment shows an attractive *incremental* return, because both alternatives have a negative present value at the minimum acceptable return.

○ A profile shows clearly situations in which the present value difference between two alternatives is small even though the incremental return appears attractive. In such cases, a decision between the two alternatives may depend more heavily than normal on the reliability of the assumptions underlying the calculation and on nonfinancial considerations.

○ The shift in the attractiveness of the alternatives for changes in some underlying assumptions may be quickly detected and measured.

Sometimes the mutually exclusive alternatives will consist of two streams of cash outlays over time without a net cash inflow. Typical examples are alternative ways of providing office accommodations, transportation facilities, bulk plant installations, and environmental protection. In such cases, as long as the need for or desirability of the function *is not* in question, the correct decision essentially rests on selecting the alternative with the *minimum* present value cost.

A cost minimization comparison for Projects C and D—replacing versus repairing existing equipment—is shown in Table 6-5. The incremental DCF return for Project C compared with Project D is 12.8 percent. Note that the net cash flow for each alternative represents

TABLE 6-5. Cost minimization comparison: cash flows for Project C versus
Project D

Year	Project C: Replace Existing Equipment	Project D: Repair Existing Equipment	Project C Minus Project D
0	(6,000)	(900)	(5,100)
1	(1,250)	(2,450)	1,200
2	(1,500)	(3,200)	1,700
3	(2,000)	(3,700)	1,700
4	(1,500)	(3,950)	2,450

the expected *after-tax cost* to the investor. Generally, when a company
is in a profit-making position, all expense outlays and depreciation
allowances may be used to offset profits from other sources. Thus,
for evaluation purposes, the actual cash outlays may be reduced by
the resulting tax credits in the development of the net cash flow.

A present value profile for Projects C and D is shown in Figure
6-9. Note that neither profile intersects the horizontal axis. As long
as it is clear that the basic objective (say, the need for truck
transportation) is worth more than the maximum present value cost

Figure 6-9. Present value profile: Project C vs. Project D

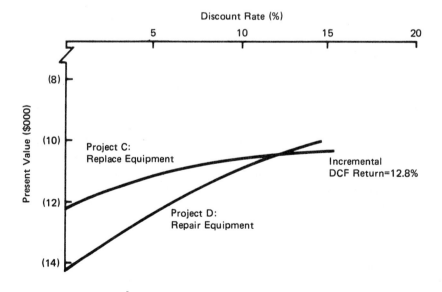

involved (just over $14,000 in Figure 6-9), either alternative will be acceptable for all values of funds. However, given a particular value of funds, one of the alternatives will be preferable.

The two lines intersect at 12.8 percent. If the minimum acceptable standard is 12.8 percent, the alternatives will be equivalent. If the minimum acceptable return is lower than 12.8 percent, Project C will be preferable, since it has a higher present value (lower cost). If the minimum acceptable return is higher than 12.8 percent, Project D will be preferable, since it has the higher present value.

If the present value curves of C and D remain close together over a range of interest rates (that is, if the present value difference between the alternatives remains small), the reliability of the assumptions underlying each alternative should be carefully evaluated before reaching a decision. It is possible that the relative position of these two lines will shift with slight changes in the assumptions.

The point of present value equivalence is referred to as the incremental DCF return realized by Project C relative to Project D. Project C is preferable when this incremental DCF return is greater than the minimum acceptable standard.

MAKING DECISIONS UNDER INFLATION

What happens to an investment prospect if inflation is expected to be 10 percent annually? Sales and operating costs are assumed to increase accordingly. Depreciation is unchanged, but taxes, profits, and thus cash flows are higher. If cash flows are higher, won't the rate of return increase? Unfortunately, no. Cash flows for each year are stated in dollars of declining purchasing power. When the cash flows are adjusted to dollars with the same purchasing power as those used to make the original investment, the rate of return decreases.

There is no change in investment or depreciation expenses, since these items are unaffected by expected future inflation (assuming the investment outlays are made over a short period of time). However, the after-tax cash flows must be deflated to dollars of a common purchasing power, again assuming a deflation factor of 10 percent yearly. The deflated or constant-dollar cash flows are then used to determine the rate of return.

Perhaps the most striking consequence of inflation is the sharp drop in the rate of return, a point developed by Brant Allen (see Figure 6-10). Even though the inflated level of sales and costs generates

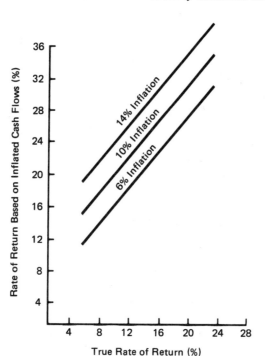

Figure 6-10.   Effect of general price level inflation on target rate of return

more cash—that is, even though cash flows are higher—the rate of return is lower. This holds for all depreciable investments made by taxpaying organizations.

The decline in the rate of return under inflation is due entirely to the income tax treatment of depreciation, which is charged against taxable income. This reduces the amount of taxes paid and as a result increases the cash flow attributable to an investment by the amount of taxes saved. But under existing tax laws depreciation expense is based on historic cost. As time goes by depreciation expense is charged to taxable income in dollars of declining purchase power; as a result, the "real" cost of the asset is not totally reflected in depreciation expense. Depreciation costs are thereby understated, and taxable income is overstated. This "inflation tax" also applies to investments that are either amortized or depleted for tax purposes (such as patents, licenses, and mineral rights) and in general to any investment that, for tax purposes, is expensed over time rather than immediately.

"Working capital drain" can also distort an investment's rate of return. Capital projects requiring increased levels of working capital suffer from inflation because additional cash must be invested at each new price level. For example, if inventory (net of payables) equal to 90 days' sales is required and if the cost of inventory increases, an additional outflow of cash will be needed to maintain this inventory level. A similar phenomenon occurs with funds committed to accounts receivable. These working capital supplements can seriously reduce a project's rate of return.

The only type of investment that does not suffer from inflation is one that can be expensed entirely and immediately for tax purposes. For example, many types of plant rearrangements, method changes, and research and development projects can be written off or expensed as incurred. If there are no depreciable assets, there can be no inflation tax; if there is no net working capital, there is no drain; and if there is no terminal sale of assets, there is no capital gains tax.

When inflation-adjusted methods are used to evaluate capital expenditure alternatives, attention must also be paid to the required earnings figure or cost of capital. To the extent that a firm's cost of capital is a function of fixed-rate obligations such as long-term debt, the effective cost of capital decreases as inflation increases. Also, to the extent that the equity cost of capital does not properly adjust to changing dollar values, the actual equity cost of capital may decrease as inflation increases. However, both interest rates on new debt and the demands of equity investors can be expected to go up as inflation increases. Therefore the cost of capital will increase with a decline in the purchasing power of the dollar. In theory, both debt and equity costs should self-adjust for inflation; in practice, they may lag significantly.

Determination of cost of capital is difficult and imprecise, even with minimal inflation. As inflation increases, the calculation becomes even more difficult. Nevertheless, if an inflation-adjusted rate of return is used to evaluate capital expenditures, the firm's calculation of capital costs must be reexamined.

One way to make the adjustment is to deflate capital costs by the dollar purchasing power index at the applicable dates. If a firm's weighted average cost of capital was 10 percent during a period when the general purchasing power of the dollar declined 8 percent, the adjusted cost of capital would be $10\% \times (1/1.08) = 9.3\%$. If projected capital costs are used, they should be deflated at the expected rate of decline in dollar purchasing power.

DCF MEASURES AND ACCOUNTING RATES OF RETURN

Analysis of a company's performance requires an understanding of the relationship between project DCF measures and the accounting rate of return (ROI). In ROI calculations, capital recovery occurs through the mechanism of an annual depreciation allowance. This allowance is not related to the annual change in the present value of future cash flows. In addition, the disposition of funds is handled without regard to the timing of the actual cash flow.

A company can be envisioned as an aggregate of a large number of projects, with corporate performance being the composite. Assume that a very simplified corporation has been constructed as a series of projects with a five-year life; one project is to be added every year. Each project has these characteristics (in thousands of dollars):

	Year 0	Year 1	Year 2	Year 3	Year 4	Year 5
Investment outlay	(500)					
Depreciation		100	100	100	100	100
Net income		34	34	34	34	34

Each project will have a 10 percent DCF return. The corporation will reach a state of equilibrium in year 5, as shown in Table 6-6. Looking at book return for the corporation (net income divided by average investment), we note considerable year-to-year variation as the corporation approaches equilibrium. The ROI (defined as the book return) levels out at 10.6 percent for the company.

At *equilibrium* the book measure of return is different from, although related to, the DCF return. However, when the corporation is not at equilibrium, the book measure can vary rather widely, even though the DCF rate is constant.

If project life is extended, the book return for a corporation will level off at correspondingly higher ROIs, as shown below:

Project Life (yrs)	Corporate Book Return (%)
5	10.6
10	11.4
15	12.2
20	12.9

For project lives of 20 years (quite typical for many industrial projects) corporate book returns overstate the project DCF return by 2.9 percentage points.

There is no fundamental reason why book return should approxi-

TABLE 66. Book data: one project added each year (in thousands of dollars)

	Year 0	Year 1	Year 2	Year 3	Year 4	Year 5	Year 6	. . .	n
Capital expenditures	500	500	500	500	500	500	500		500
Gross investment (beginning of year)		500	1,000	1,500	2,000	2,500	2,500		2,500
Net investment (beginning of year)		500	400 500	300 400 500	200 300 400 500	100 200 300 400 500	100 200 300 400 500		
Beginning-year total	0	500	900	1,200	1,400	1,500	1,500		150
Average-year total	250	700	1,050	1,300	1,450	1,500	1,500		1,500
Net income		32	64	96	128	160	160		319
Book return: (%) net income/average investment		4.6	6.1	7.4	8.8	10.6	10.6		10.6
Depreciation flow		100	200	300	400	500	500		1,000
Cash inflow from projects		132	264	396	528	660	600		660

mate DCF return. In simplified terms, the numerator for the ROI fraction contains only (after-tax) profit; for DCF it includes both profit and depreciation. In analyzing investments, company managements are more likely to use the conventional ROI, calculated with *total* capital investment as the denominator (rather than average capital investment); in that case the ROI will be a considerably smaller percentage than the DCF return.

We have discussed how a DCF return can be calculated for any set of cash flows received over a period. Obviously, when capital investments are involved in the cash flows, we can also calculate an ROI. If the cash flows vary year by year, so will the ROI. Indeed, the ROI generally increases slowly (see Chapter 7) and may or may not attain a constant value.

For the simple case of constant cash flows, the ROI and DCF return can easily be related. This is shown in Figure 6-11 for projects having depreciation schedules of 5, 15, and 25 years. The relationship could be embellished by considering nonuniform levels of cash, but this is not in our judgment worthwhile.

Figure 6-11. ROI vs. DCF return

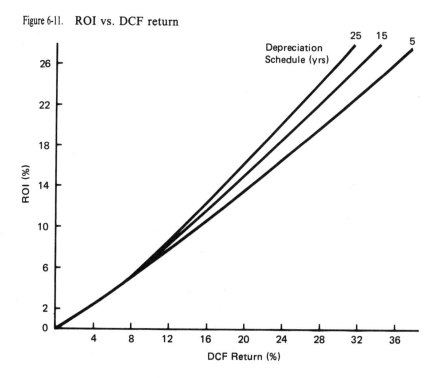

chapter 7

The Elements
of a
Venture Analysis

All of us come up, at one time or another, with ideas on how our businesses can be improved. Some of these ideas may even be good. Many, however, die in their infancy because we don't launch them properly. The diversity of commercial opportunities arising from an imaginative research program creates a unique set of management problems. The most common problem involves acquiring and assessing in a uniform manner the important facts on which management decisions will be based. In a large, decentralized company the problem can be compounded. What is needed is a common framework for organizing and processing information to guide business decisions. Venture analysis, which builds on concepts we have discussed in earlier chapters, provides such a framework.

Assume you are working in your company's exploratory plastics group (whatever that may mean to you). Your job is to come up with product modifications that will extend your company's product line or, happy day, to discover a new plastic. You have a good technical background in the chemistry and engineering involved in your product area and are experienced in the fabrication industry using your company's products to manufacture their products. Technically skillful though you are, you are unsure of how to go about getting a good idea off the launching pad.

In the course of your work assume you have come up with a new plastics resin, put together from some combination of raw materials, catalysts, and process operating conditions. Your laboratory results indicate a set of physical properties that seem to offer an advantage over the commercially available products you are aware of. You visualize that the availability of a plastics resin with these properties will enable several fabricated products to be switched over to your product *if the price is right.*

Small laboratory-scale samples sent out selectively have elicited enthusiastic responses from several companies representing a spectrum of different end-use applications. These firms are potential customers. The total feedback, both external and internal, suggests that you have a "winner" on your hands, and you are getting excited.

You are aware of a multitude of technical and marketing considerations that require further investigation. Processing refinements and optimizations need to be worked on. The sensitivity of product properties to process changes has to be explored. The potential sales volumes and prices have to be restudied often, and at a later stage marketing strategies and tactics must be evolved. Ultimately process designs and estimates of investments and production costs will be required for the prototype and, subsequently, commercial plants.

But before all these efforts are put in motion, you must begin a serious economic analysis. Before too many dollars are spent on the presumption that you have a winner, you must convince yourself and those who run your company that a winner is what you have.

Your management has a responsibility to see that healthy performance continues on the bottom line of your company's profit-and-loss statement. That responsibility means that commercial implementation of your idea will have to meet some minimum financial criteria. In addition, your idea will compete with those of others in your organization. Therefore, like it or not, you have to put information about your project in economic terms.

What, then, do you have to create? In earlier days, when economic concepts were simpler and easier to live with, you might have gotten by with showing management that it would cost so many dollars to build the plant and that the plant would produce *x* pounds of product annually. The product would cost *y* cents per pound and would be sold at *z* cents per pound. When this plant was producing all out, the profits after taxes would generate a "conventional return" ratio of net income to investment of "such and such" percent. You might have even added the bonus of calculating the payout period, or the time required to get the investment back.

Nowadays, managements are more demanding, if only because of past mistakes. They are interested in knowing about the total capital resource requirements of the project, which will include working capital in addition to the fixed capital cost of the plant itself. They know that the discounted cash flow (DCF) return calculation is superior, since it recognizes the time value of money and thus facilitates choices among alternative courses of action.

Managements will also want to see a cash commitment (impairment) profile, which establishes the size and timing of the funds at risk. Not only will they require these details about the specific project you have in mind; they will also want to know about the expanding picture you foresee beyond the initial project. In this case, you will have to come to grips with creating a *venture analysis*. Unquestionably, not everyone is going to agree totally with every assumption you build into your calculations. However, you must be perceptive enough to anticipate those areas where challenges are likely to occur and prepare a *sensitivity analysis* beforehand. In doing so, you must be able to cope with questions on how things would change if different prices, sales volumes, or investments prevailed.

Eventually, in developing your project (process or product) you are going to have to cope with all these things, but many of them can wait until later. For now, you need to make estimates of timing, market size, investments, working capital, and raw materials and operating costs; you must also take into account certain tax effects.

TIME INTERVAL FOR INNOVATION

In earlier chapters we examined the pattern of innovation and considered some rules of thumb about the costs of commercializing an idea. Our conclusion was that after an invention has been made and R&D completed, the route to innovation is long and costly.

J. Jewkes, D. Sowers, and R. Stillman, in their book *Sources of Invention*, collected data on the lag between invention and successful introduction to market for major chemical, mechanical, and electrical innovations. Although there is no systematic chronological pattern, less than half the major chemical innovations of the twentieth century show a time interval of less than ten years between patent issuance and first introduction to market. (Since activity is likely to continue between the invention and patent issuance, the time interval is probably understated.) For major electrical and mechanical innovations, the average time lag was more than ten years (see Table 7-1).

All indications are that this time lag is increasing rather than shrinking.

In Chapter 1 we introduced the concept of a project planning matrix as part of the discussion of innovation. Charlie Stokes has followed the development of three major chemical innovations. The planning matrix for each of these is given in Figure 7-1. Note the

TABLE 7-1. Time interval between invention and first commercial introduction

Innovation	Date of Invention	First Commercial Introduction	Interval in Years
Chemical Inventions			
Crease-resistant fabrics	1918	1932	14
Nylon	1928	1939	11
Silicones	1904	1943	39
Teflon	1941	1943	2
Penicillin	1928	1944	16
Streptomycin	1939	1944	5
Polyethylene	1933	1944	11
Orlon	1942	1948	6
Polyester fiber	1941	1953	12
Mechanical and Electrical Inventions			
Power steering	1925	1931	6
Helicopter	1909	1932	23
Radar	1922	1935	13
Television	1919	1941	22
Jet engine	1929	1943	14
Ballpoint pen	1938	1944	6
Long-playing record	1945	1948	3
Xerography	1937	1950	13

time span to commercial market introduction (plant startup). The critical question is: What is the *time span for your idea* likely to be? Time span depends not just on the nature of the idea and how it will be used but on the organizational approach (or desire). However, for a major idea time spans of less than four or five years are unlikely.

Sensitivity analysis (discussed below) will almost always show that a reduction in the time span improves the attractiveness of the

Figure 7-1(a). Planning matrix for a family of entirely new pharmecenticals

Figure 7-1(b). Approximate planning matrix: Polybutadiene development

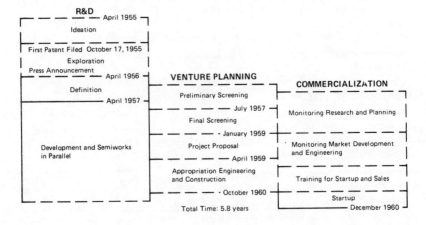

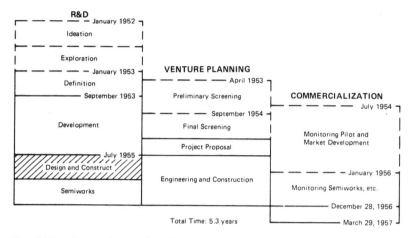

Figure 7-1(c). Approximate planning matrix: High-density polyethylene development

venture, sometimes significantly. The key is to reduce the time span without introducing some serious debits in the process. Optimism is insufficient.

_____ MARKET

The information pouring in to R&D managers about market size and price—whether for new products or established products—is increasing in kind and quantity. To evaluate market data, managers will have to depend on their experience and skill. Their objective in studying the market is to develop two models:

Market model—a sectioning of the market into appropriate segments and development of information about each segment as a function of product price and customer value.
Marketing model—a plan for penetration into the market, taking into account competitors' actions, products, and prices and their potential influence on customers.

Note the distinction between projecting the size of a market that is open to a product and determining how sales are going to come about.

If you are considering or conducting R&D on one of your company's established products, the market and marketing information

should be available to you, but it will probably not be in the form
you desire. You are going to have to do quite a bit of work to put
information into a usable quantitative form.

Consider, for example, the data on low-density polyethylene
production shown in Figure 7-2. (Note that production has been
normalized by GNP to aid in extrapolation.) If the venture analysis
will be carried out with computer software, or using the market
specification approach (see Chapter 10), there are many benefits to
fitting the data with an empirical expression such as the following:

Figure 7-2.   Demand for low-density polyethylene

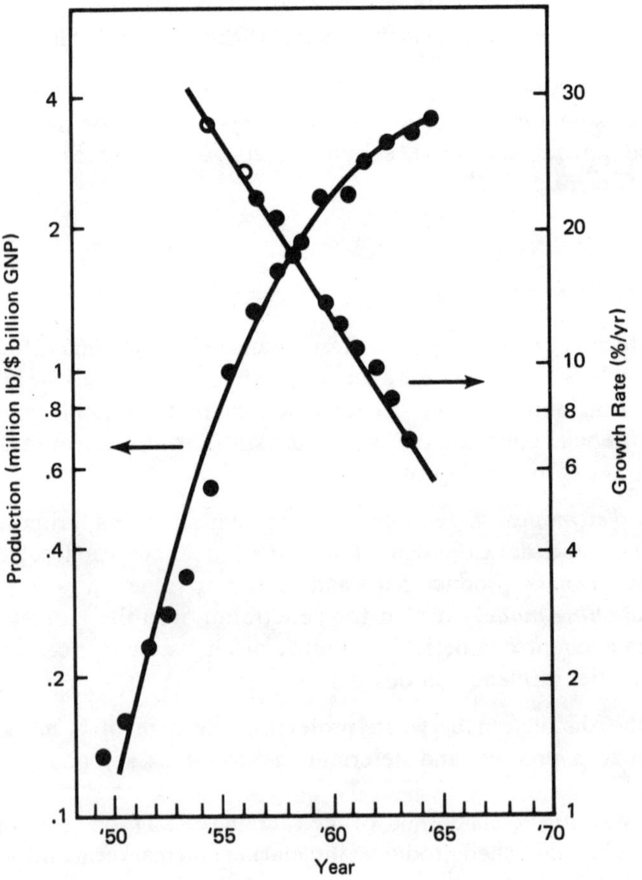

$$S_t = S_0 \ [e^{r_0(1-e^{-.176t})/.176}] \ \frac{GNP}{GNP_0}$$

where S_t = sales volume in year t
S_0 = sales volume in 1949
t = time in years
r_0 = initial rate of sales growth (about 40 percent per year)
e = base for natural logarithms

Other expressions can undoubtedly be fitted to the data.

Your sources for market information on the new plastic are limited. Obviously you are interested in obtaining information about the size of the market for which the product is technically adequate. You will use more than one technique to get crosschecks. The possibilities include:

Opinion polls. Although polls are subjective, they can give you a good feeling for the potential market. We have found telephone polling to be quite effective.

Historical analogies. If you can find a similar product (and are willing to live with the idea of a similar business environment), you can generate some excellent projections quickly. You can be a little elastic about what "similar" means. The literature in this area often uses the nylon–polyester analogy, but that's a bit too restrictive.

Simple trend projections. Use common sense, not technology diffusion models. Be sure to compare the projection with your company's experience. Remember what happens if you assume a growth rate of 20 percent per year for a decade.

Correlation analysis. GNP is the most popular correlation variable, although many others may be better—for example, the Federal Reserve Board's Index of Durable Manufactures or some statistic(s) from the industries your product is targeted for.

Mathematical models. Most of these start from input–output analysis, although there are other possibilities. They can become too complex if you are not careful.

In predicting markets, you can get much help within your own company, of course. Your company's market research group will probably have a huge store of background data, much of it purchased from consulting firms such as SRI, which offers the 20-odd-volume continuously updated *Chemical Economics Handbook*. Chemicals company market researchers usually spend most of their time forecasting markets for new products (new, at least, to the company) rather than existing products. Initially their forecasts may involve extrapolation of past production trends. As interest deepens, they shift to projections of end-use markets and possible penetration by the new product. The industry's market researchers belong to a professional organization called the Chemical Marketing Research Association (CMRA), about 800 strong, which holds four meetings a year. Each meeting yields up to 20 papers on projections of markets and products; all are published and available. More specialized groups—plastics market researchers, agricultural chemicals market researchers—have split away and formed their own groups, which also meet regularly every year.

Whatever you do, keep it simple and keep it current. It won't be current if it isn't simple, and complexity is often not justified.

With the market model in hand, you must develop the marketing model; that is, you must decide how to get those sales (remember, sales are caused), what programs are envisioned, and so on. You must also determine how to set expenditure levels. During the market entry or pioneering phase (depending on which version of the product life cycle you use) there is no limit to the money you can spend except your ingenuity, the available manpower, and management restraints. The critical question is whether all that money and effort will position you for takeoff.

If you have done all that is implied in these short comments (you will find the Notes on this chapter valuable for additional information), you will have as an end product the analysis portrayed in Figure 7-3. Your market model studies will yield estimates of the total sales potential or opportunity for your product. This market may be yours exclusively for a period of time, or it may have to be shared with competitors. In preparing the model, you will have to consider your position (share) in the market.

_____ SALES

Sales revenue is the predicted future annual volume of products sold to customers multiplied by the predicted future price per unit of volume. Other terms having the same meaning are "income" (on the statement of income and retained earnings), "net sales" (on the operating and revenue statement, where "net" recognizes that product is sometimes returned or adjustments are made), "Net income from sales" (abbre-

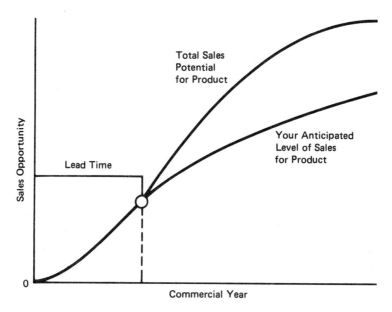

Figure 7-3.   Sales opportunity model

viated NIFS), and "sales income." Revenue may also be referred to as "additional sales" in order to distinguish between the income that is made possible by a project and the income that would have occurred in the same business area if the project had not been executed.

Accurate prediction is impossible because the act of selling must be accompanied by the act of buying, and buying is subject to decisions not under the control of the seller. Buying decisions are, first and last, made to fill the buyer's need, which may range from a need for prestige to a need to make a profit on the further processing of the purchased commodity.

The only way that a sales effort can be successful, or that a sales prediction can become reality, is to determine the buyer's need and to provide a product that satisfies it at a competitive price the buyer is willing to pay. This is a tall order, and it depends greatly on the skill, perception, diligence, and intuition of those who deal with the buyer. But to no less an extent it depends on the skill of those responsible for creating a high-quality product at minimum cost. The seller also has a need: to make a profit. Therefore, the seller must make a satisfactory profit at a price the buyer will pay.

The setting of a sales price proceeds through three stages. First, the volume needed by the buyer is predicted by *assuming* that some price acceptable to both buyer and seller can be offered. An acceptable price does not create a need for a product. It only enables a buyer to satisfy a need that already exists. Second, the price that will yield a satisfactory profit to the seller is calculated from the predicted volume. This is the lowest price, or floor, below which no offer to sell should be made. To this end, the seller must consider reasonable criteria for profit that reflect both the specific business area and the company. Third, armed with some knowledge of the effect of price on profit, the seller must decide on a price that the buyer will willingly pay. Obviously, this is not straightforward.

An analysis in support of these three basic steps will sometimes supply additional information. Inasmuch as *our* profit is affected by the price we pay for commodities, an analysis of a *supplier's* profit may reveal that the company could stabilize or lower its costs. Likewise, an analysis of the profit of *competing producers* will show whether the company can profitably match or beat the likely competitive price. Finally, an analysis of the *customer's profit* will establish the "ceiling" price that the company can ask.

Specific analyses are beyond the scope of this discussion, since all suppliers, competitors, or customers have their own way of manipulating their incomes, costs, investments, and criteria into statements of profitability. We could guess fairly well at their cash flows if we knew what their estimating procedures were, but we could not be sure how they defined cash flows or what they considered to be an adequate criterion of profit. Second-guessing is more of an art than a science. However, rest assured that while we are second-guessing outsiders, they are second-guessing us.

In any analysis of the effect of price on profit, the interrelationship between price and volume must be considered. As stated before, lowering the price of a commodity does not create a need for the commodity. However, it may well permit buyers to satisfy needs that would have remained unsatisfied at a higher price or that would have been satisfied with a less desirable commodity. In almost all cases, the price must decrease if the volume is to increase significantly. At the same time, the unit costs will decrease as the volume increases, although perhaps at a slower rate than the decrease in price. Thus, up to the "point of diminishing returns," an increase in volume will yield a greater income.

From this it might be inferred that there is no reasonable limit to the volume that should be sought, given a price structure that is not too sensitive. Barring legal and ethical considerations, this is a good working premise, provided that the fixed-investment requirements have also been taken into account.

An important concomitant of the price–cost–volume relationship is the investment–volume relationship. One reason why fixed investment is so important in pricing decisions is that the character of the investment may change with volume. Although the chemistry of a process is not likely to change with increasing volume, significant physical changes usually take place. That is, the "hardware" is different, and consequently the investment per unit of product is different. For example, the production of only 100,000 pounds per year of an average chemical product would almost certainly be conducted batchwise in equipment that had a much greater capacity. In order to take advantage of the spare capacity for other products, the equipment would be more elaborate and costly than required solely for the product in question.

However, 100 million pounds per year would be produced continuously in equipment designed optimally for the single product. Thus as a general rule the fixed investment per unit of product, as well as the manufacturing cost per unit of product, decreases with an increase in volume. The allowable price floor is therefore lower at the higher volume. To the extent that a lower price can induce buyers to satisfy more of their needs with the product, knowledge of the interrelationships among price, cost, investment, and volume is a powerful weapon in the hands of management.

The classical situation requiring a pricing decision occurs when the long-range volume has been predicted within reasonable limits. (Reasonable limits might be defined as a two-to-one ratio of the maximum likely volume to the minimum likely volume.) In this situation it is usually possible to design an efficient facility with a capacity equal to or slightly exceeding the most likely volume and to compute accurately the costs and investments for all levels of operation. There are then only two variables whose effect on profitability must be studied closely: price and level of operation. A graphic representation is useful in such cases.

If the required value of a chosen profitability criterion has been decided (say, 20 percent ROI), a graph similar to Figure 7-4*a* may be plotted. If the required value of a chosen profitability criterion (such as ROI or DCF) has not been decided upon, the criterion may be shown as a variable, as in Figure 7-4*b*. If the price is fixed but

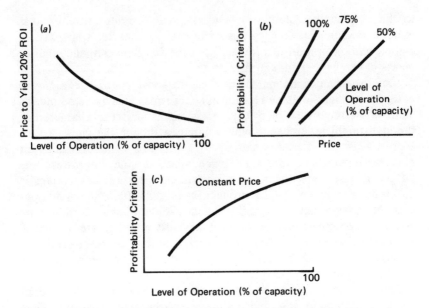

Figure 7-4.   Effect of price and level of operation on profits. (a) Required value of chosen criterion already decided (ROI). (b) Criterion as a variable. (c) Price fixed, but required value of profitability criterion undecided

the required value of the profitability criterion has not been decided, the graph in Figure 7-4c is customary.

In graphs such as these there are many ways to position the variables. In general, it is customary to place the criterion on the *y* axis if it is a variable. If the criterion is not a variable, the *y* axis should be used for the dependent variable (price in a marketing analysis).

The importance of the market study is indicated by an analysis prepared by Chaplin Tyler of Du Pont of 12 projects that ran into difficulties. These cases represented investments over $200 million. The lower than expected performance was caused by:

Failure to attain forecast sales volume in three cases.

Lower than expected selling prices in one case.

A combination of lower sales volume and lower selling price in three cases.

For 7 of the 12 failures the cause was lower than anticipated sales, a situation that typifies the difficulties encountered in hundreds of projects reviewed by Tyler over a period of 15 years. The solution is not necessarily a better market survey; in many cases the survey is the best that can be made under existing conditions. Thus project analysts cannot dismiss their responsibilities in this area so lightly. We shall discuss later just how sensitivity analysis can be used to broaden the basis for management judgment.

Another important consideration in an analysis is knowing how revenues have been determined. When a manufacturing plant does not handle its own distribution, the *netback*, or revenues actually realized by the plant, may be used. The difference between netback and the selling price of the product is that it takes into account selling and distribution costs, which can be high. Therefore, if we use actual market prices in our evaluation, we must also include selling expenses and the cost of transporting the product to market in the costs.

Also, only under unusual circumstances will capacity output be attained in the first year after startup. More frequently, two, three, four, or even more years are required, particularly in new product ventures. On the other hand, realizations may be lower in later years because of competitive pressures, gradual removal of duty protection, or any number of other factors. For these reasons, we should consider both a short-range and a long-range price structure for both product and raw materials costs in any analysis. Year-to-year variations in the pattern of sales volumes and realizations must also be considered.

NEW FIXED INVESTMENTS

Investment is synonymous with asset. It follows that fixed investment has the same meaning as fixed asset on a corporate balance sheet. More particularly, fixed investment is taken to mean fixed assets at cost, with "cost" being the acquisition cost before deducting accumulated depreciation allowances. The acquisition cost of fixed investment is also referred to as gross book value. Gross book value less accumulated depreciation is referred to as net book value, or simply book value. In order to indicate the commitment of monetary resources before depreciation takes place, both the ROI method and the DCF method define fixed investment in terms of gross book value or acquisition cost. Depreciation allowances are calculated separately.

In accounting terminology, the transformation of expenditures into fixed investment is called capitalization, and the expenditures are said to have been capitalized. The distinction between capital expenditures and noncapital expenditures is an extremely important one, in part because of the effect on income taxes. If a capitalizable expenditure is treated improperly as noncapital (that is, as an operating cost), a corporation will underpay its taxes—with obvious consequences from the Internal Revenue Service. In addition, an understatement of assets will probably have an undesirable effect on the value of a corporation's stock.

Typical practice is to capitalize any "real" asset of a permanent nature intended for long, continued possession or use in the business. Practical considerations suggest that an item should cost $100 to $500 before it is treated as a capital addition. When a new plant or a major addition to an existing plant is constructed or equipped, *all* machinery, equipment, furniture, and fixtures are capitalized regardless of length of life or value. This has been a good working definition for many years, although the number of special interpretations would undoubtedly fill several volumes. The definition is, after all, a reflection of accounting and legal principles, and it is quite safe to say that the last word has not yet been written.

Normally a small new company that is barely breaking even would prefer to capitalize as many dollar outlays as possible to be able to show some profit to stockholders and potential investors. A large, profitable company would prefer to expense all dollar outlays to save taxes. For years, for example, arguments have been waged over the accounting (tax) treatment of oil wells drilled by petroleum companies. The rules now permit dry holes to be expensed, but producing wells must be capitalized. At every session of Congress legislation is proposed to force the oil companies to capitalize all drilling expenditures and thereby pay more taxes. For some years, to stimulate capital investment (and employment), the government has given companies a financial incentive to build more new plants, sooner. This "investment tax credit" has recently been 6 percent of the capital investment. The credit in effect allows companies to double-depreciate part of their investment in new plants. Currently, the government is considering increasing the investment tax credit as well as permitting companies to accelerate (shorten) their depreciation schedules.

As an illustration of the distinction between capital and noncapital expenditures, we can refer again to our hypothetical plastics project. This will also be a useful introduction to the many items that must be taken into account in preparing an estimate of the physical requirements (equipment, buildings, and so forth) for a new project. Assume that the new plastic has been perfected and is to be manufactured in a new plant. In the beginning, estimates of sales income, operating costs, and investments were made, based in part on the historical cash flows in similar projects. The estimates were submitted to management, which approved the project and authorized the expenditure of funds.

The man-hours spent in making the estimates and preparing the appropriation proposal were noted but were not included in the project cost and were therefore not capitalized. Immediately after appropriation, the project was divided into more or less independent components: a reaction system, a distillation system, a control building, and an equipment structure. The expenditure for each component was estimated. One purpose of thus subdividing the project was to gain more accurate accounting control of subsequent expenditures.

Many estimates of investment are based on something less than detailed designs and layouts. Therefore, it was necessary to complete the engineering studies, whose cost was also capitalized. After the major items of equipment were designed and their approximate location agreed on, a scale model of the new plant was constructed. This model, whose cost was also added to the project, was used to develop detailed drawings for foundations, piping, wiring, sewers, pavements, fire protection, and structural steel.

Various purchasing groups then began to negotiate the prices and to confirm the specifications for the major items of equipment, as well as for the material that would go into the foundations, piping, and so on. When the negotiations were completed and were approved by the groups that had made the designs, contracts with vendors were drawn and signed.

In the meantime, work was proceeding in the field. Earth-moving equipment and construction tools were assembled. Temporary buildings were erected to serve construction personnel. Ground was leveled and excavated for underground lines and foundations. Structural steel—some left over from previous projects and some expedited at a cost premium for the project—was erected. Equipment foundations and dikes were poured. Sewers and drains were installed and backfilled. Taps were made into the main utility headers and electrical conduits.

As the items of equipment began to arrive, they were unloaded, uncrated, inspected, tested, and tagged with item numbers. As they were needed, they were brought to the construction site, unloaded, and installed. The facility was insulated and painted. Lights were installed, access roadways were built, concrete operating pads were poured, gravel was spread, and grass was planted in appropriate areas. The process equipment was given a final inspection and all vessels were closed. Motorized equipment was lubricated, turned on, and adjusted. All debris was removed from the area. Finally, operating personnel were notified that the facility was ready to begin production.

Throughout this period all expenditures were carefully accounted. When operating personnel were satisfied that the installation was complete and in perfect mechanical condition, the work orders were closed. No further charges for rearrangements or repairs were accepted on the capital orders. The total costs for each order, and for each item on the capital order, were then derived.

The derivation of subtotals and item totals is difficult. Although the early expenditures for purchasing and design had been carefully noted, it was impossible to delineate all the time charges to individual items, or in some cases to individual categories. The difficulties also extended into the field, where accounting personnel could not be sure how much time each painter, welder, electrician, and mechanic spent on each item. However, recognizing these difficulties in advance, the accounting department set up accounts for collecting nonspecific charges and obtained agreement on logical methods for allocating these charges. The systematic collection and allocation of charges is done through cost of accounts—a logical pigeonhole for related charges.

There are at least five reasons why the project costs had to be broken down into items or groups of items:

Cost control of a project is improved by focusing on small segments.

Except for intangibles such as riparian rights, fixed assets must represent tangible objects and all labor charges must be attributed to such objects.

Procedures governing depreciation specify different rates for different classes of assets; therefore, they must be valued separately.

Cost correlations encountered in a project help improve the accuracy of subsequent estimates.

Assignment of investments to products is aided if the investments are subdivided.

In our example, major equipment (converters, columns, pumps, heat exchangers) formed a natural group of cost items. There were, however, several large expenditures (structural steel, concrete pads, sewers, roadways) that could hardly be called equipment, as well as a multitude of small physical objects and not-so-small labor man-hours (excavation, painting, insulation) that could not be attributed to any specific item of major equipment. Therefore, the "cost of accounts" approach was used to identify and group expenditures in a more practical fashion: major equipment items, buildings, steel structures, property improvements, piping, wiring, instrumentation.

So far, the project costs do not represent depreciable assets. While the expenditures were being made, they were known as "property in the course of construction," and one more step was necessary. By a process called "clearing," the appropriate financial group in each manufacturing location was informed of the acquisition cost, acquisition date, and date of placement in service, and was given a description of each item of property. Only when this information was entered on the company's property account books did the costs achieve legal status as depreciable fixed assets.

The estimation of fixed investment depends, therefore, on what expenditures are customarily capitalized. As stated before, no single rule applies; but in general any expenditure directly involved in the creation of physical facilities—up to but not including initial operations with raw materials—is capitalized. Excluded are process raw materials, catalysts, heat transfer fluids, removal of existing assets, and training of operating personnel, even though these expenditures may occur during construction. The expenditures for a processing facility may be summarized as follows:

1. Equipment purchase price.
2. Labor for installation.
3. Miscellaneous material for installation.
4. Pipe racks (estimated independently and broken down by materials and labor or estimated as total subcontract):

Piles	Power wiring
Foundations	Equipment lighting
Trenches and pits	Fire protection
Piping	Painting
Sewers	Spare parts
Instrumentation	

5. Expense (represents costs of purchasing, expediting, inspection, and maintenance of field forces).
6. Engineering (represents cost of detailed designs and specifications).

7. Contingencies (represents a safety factor for unforeseen costs other than changes in scope).

Any existing assets on the site of the new facility will probably have to be removed, relocated, or rearranged in some way. The capitalized value of these existing assets includes the costs of such things as design, layout studies, and installation; thus part of the value to which those costs contributed disappears after the existing assets are removed. In effect, part of the fixed assets of the company is abandoned or "retired." It is not necessary to quantify the amount of retirement in an economic analysis, but it may be necessary to include the salvage value (as an income) if it is large, or to include the effect of large depreciation writeoffs.

New investment is, of course, investment that is proposed but not yet made. Another distinction is not quite so clear: new investment is generally limited to expenditures that represent the peculiar needs of the project in question. In this sense, it is sometimes called direct investment. Recall, for example, that during construction of the new plant taps were made into the main utility headers and electrical conduits. This implies that the project will utilize "supporting" investments in facilities for the generation and distribution of steam, water, electric power, and other utilities. If spare capacity exists in these utilities, no new fixed investment in utilities will be required. However, if one or more of these utilities are in short supply, the project may require installation of a new boiler at the powerhouse or a new pump at the cooling-water intake valve. Even if a new boiler or a new pump is required, it may not be included in the new fixed investment for the processing facility.

Let us delay for a moment the question of how the investment in supporting facilities, either new or existing, is portrayed in an economic analysis and turn instead to our estimate of investment for a processing unit. We need to make estimates of the following: onsites (the manufacturing unit itself), offsites (related facilities, often used jointly with other units), utilities, tankage, and supporting facilities. In order of decreasing accuracy, the options open to use are:

Detailed engineering estimates (there generally are a series of these that improve in accuracy).

Estimates made from data (or so-called investment curves) based on comparable projects.

Contractor's study estimates.

Investment information from the literature.

The most reliable estimate is made during the equipment procurement and erection phase of a project, since mechanical design of vessels, piping layouts, foundations, and structures is generally well under way or completed. This type of estimate relies on detailed computation and includes, to the extent available at the time, firm bids or vendor quotations on major equipment and costs for bulk materials, engineering, labor, and hardware based on specific quantities and unit prices defined by detailed engineering. Actual financial commitments up to the point of the estimate are also included. This type of estimate has a high degree of accuracy: actual costs are likely to be within 10 percent of the estimate.

There is no completely satisfactory method for making investment estimates at the start of a project. All known methods are either too expensive or too slow, require too much computer time, or are too crude. However, people have learned to live with the problem, and many techniques (all compromises of one sort or another) have been developed.

The most accurate evaluation of investment requirements for a new process is based on firm engineering studies. In the early stages of a new process, when many development decisions are made, available data are simply inadequate to prepare this type of estimate. Moreover, some cynics have suggested that regardless of the state of the technology, the actual investment will be known only after the plant has been built.

Lower-cost, more rapid methods deal with equipment groupings and systems and are based on the application of factors to the purchase price of major equipment. These range from the traditional Lang method—which uses a single factor, depending on whether solids, liquids, gases, or a combination of these is being handled—to the rather detailed factoring system described by Francis Zevnik and Bob Buchanan. Our experience has been that the methods suggested by Zevnik and Buchanan provide investment estimates acceptable for the venture studies we have in mind. If computer capability is not available in house, several firms provide versatile time-shared computer software that can be used even with preliminary flowsheets.

But what if you have even less than a simple set of engineering drawings? Then Figure 7-5, which is based on our recent experience, may be of some help at the start. Unfortunately, as cost escalation

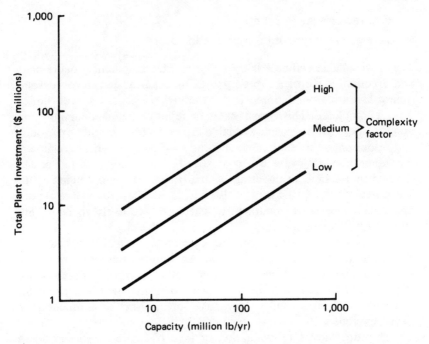

Figure 7-5. Plant investments: 1976 cost basis, where offsite investment equals 45 percent of onsite investment

continues, you will have to update the curves with the data from your own experience. Simple corrections by prorations with a cost index are not sufficient.

What does the single-complexity factor "low, medium, high" in Figure 7-5 mean?

Low—batch processing and simple synthesis (NH_3, MeOH).

Medium—continuous processes involving only gases and liquids at modest temperatures and pressures.

High—processes involving significant solids-handling polymer units, high-pressure and/or high-temperature operations.

Obviously, the curve is only a crude approximation, but it is a starting point. (Its two Σ, 95 percent confidence limits have been judged to be accurate to $\pm$ 30 to 40 percent for a large population of chemicals plants.)

_____ SUPPORTING FACILITIES

If every operation in a company were completely separate and self-contained, there would be no supporting facilities. Each manufacturing operation would buy its own land, manufacture or purchase its own raw materials, generate or purchase its own utilities, keep its own books, pay its own taxes, meet its own payroll, operate its own laboratory and dispensary, distribute and market its own products, and maintain its own supply of capital. Many manufacturing operations would not show a satisfactory profit under such conditions.

Supporting facilities begin at a place called the *battery limit* of the facility in question. The battery limit varies with the kind of supporting investment. In the case of steam, the battery limit for a processing facility is just downstream of the takeoff valve on the main pipe rack. In the case of cooling water, the battery limit is just downstream of the main takeoff valve, wherever it is located. In the case of electric power, the battery limit is just downstream of the substation that serves the facility. And in the case of sewers, the battery limit is just downstream of the manhole at the plant roadway. Investments outside the battery limits, although utilized by the facility, are not peculiar to the needs of the facility. Other facilities could also use these investments and probably will if the facility in question does not.

A multitude of supporting functions do not increase in cost in direct proportion to the amount of support rendered. For that reason, it is good business to consolidate a number of "customers" so that they may be served collectively by large, more efficient supporting functions. However, the size of many supporting functions does increase somewhat when greater demand is placed on them. The increase may not happen immediately, but it is certain to happen sooner or later unless the supporting function is large enough to handle all future requirements. Therefore, every prospective new "customer," every proposed new project, must justify the future increases in supporting-function costs and investments for which it shares responsibility. The object of a long-range incremental investment analysis is to take into account the future economic effect of a project on the fixed investment in a shared supporting function.

Before progressing further, we should point out that support is by no means limited to manufacturing operations and is by no means a one-way street. With few exceptions, every corporate function receives support from and gives support to other functions. The payroll department gives support to the laboratory and receives support from the utility and maintenance departments. The steam generation department gives support to the cooling-water department, which reciprocates with support for the steam generation department. One manufacturing operation gives support to other manufacturing operations that convert its products into derivatives. Therefore, few economic analyses can escape evaluation of a supporting facility.

As a concrete example of long-range incremental investment, we can look at an actual project that involved the addition of a small facility to a large complex. The numbers (in thousands of dollars) looked like this:

Investment	Current Cost	Long-Range Cost
Onsites	8,100	8,100
Offsites	1,300	2,100
Utilities	600	2,500
Tankage	2,200	2,500
	12,200	15,200

The difference between offsites and utilities components lies in the fact that on a current-cost basis, offsites do not include a pro rata share of buildings and other site items such as sewers, fences, and grading. In addition, it may be assumed that most nonsupervisory personnel will come from existing site forces and that the existing site layout is adequate. Current-cost utilities do not include investment for power and steam generation. On a long-range-cost basis, utilities do include power and steam generation.

Thus, in general, the investment used in developing project economics must include not only the total erected cost of the actual equipment required but also a pro rata share of the cost of any existing supporting facilities that the new plant uses. There will, of course, be some situations where the long-range incremental concept does not apply.

Land can also be a long-range incremental investment cost. For example, if we install a new plant we must include piping, storage, and loading facilities for getting the products out of the plant. However, in addition to these offsites, which must go in immediately, our new plant may require a fenced-in, graded, and sewered area, an administration building, a cafeteria, shop space, and utility capacity. Assuming that the new plant is being integrated into an existing complex, these facilities may already be available in part or in total.

Generally, supporting facilities are expanded at regular intervals rather than a little each time a process unit is added. They are usually a little oversized to begin with and are then pushed to capacity or slightly beyond before another expansion is undertaken. For instance, in an expanding manufacturing site, steam generation facilities are normally installed in reasonable-sized increments, say 50,000-pound-per-hour boilers. A new project at this site would first make use of existing available steam capacity. If an additional 10,000 pounds per hour of steam were required, it would probably be added as part of a new 50,000-pound-per-hour boiler rather than as a new 10,000-pound-per-hour boiler. The project is charged only with the incremental capacity of 10,000 pounds per hour, not the full cost of the 50,000-pound-per-hour boiler. The difference between the two represents the company's preinvestment in steam generation facilities. In effect, as several projects requiring small amounts of steam are added to the site, each project does not include its own small, separate steam plant; rather, all the projects share the costs of a larger, more economical steam plant, and they are charged accordingly.

But the fact remains that even if a project can, for the present, make use of existing facilities, in the future some of the existing offsite facilities will have to be expanded because of the addition of projects to the complex. Therefore, a new plant should be charged with the investment in the associated facilities it uses. Since this change must be reflected in the economic analysis, a long-range incremental approach is needed.

Let's consider for a moment the consequences of not using a long-range incremental approach. Assume that there is just enough spare steam capacity at the boiler house, waste disposal capacity, and room in the office building to handle the addition of a project. Then what happens to the new small addition that exceeds the available boiler capacity? It will have to bear the burden of a new boiler house. The return on an individual piece of equipment will rarely be able to support a new boiler house, a new cooling-water main, a new office building, and so on. Therefore, it may be hard to justify projects in the future, unless a real star comes along. Hence, if we were guided strictly by out-of-pocket facilities, future expansion would be difficult if not impossible.

Quantification of a long-range supporting facility is based on replacement cost and is charged to the project in the future years when additional capacity will be needed as a result of the decision to proceed with the project. This principle is almost axiomatic; it recognizes that it is the future that counts and that the original costs of the supporting facilities utilized by a project are not applicable except insofar as they are a guide to what the replacement costs may be.

Like any principle, this one requires a certain amount of interpretation in order to be useful. To begin with, consider "replacement cost." The replacement cost of a facility is hardly ever the physical duplication cost. As long as technological and architectural changes continue to be made, it is unlikely that any existing facility will be physically duplicated. Rather, it is the *function* of the facility that requires duplication, or replacement.

The situation we face is summarized in Figure 7-6. Three items of data are needed before the figure can be useful: capacity of the supporting facility as it presently exists; size of each incremental expansion that would occur if and when expansion is necessary; and cost of each incremental expansion. Imagine that the existing capacity is greater than the highest point of either curve in the figure. In

Figure 7-6. Impact of a project on supporting facilities

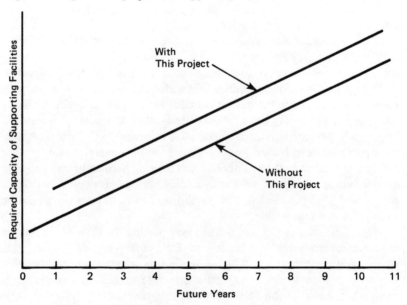

that case, *no* foreseeable additional capacity will be needed as a result of the decision to proceed with the project. Therefore, no long-range investments for this particular supporting facility should be accounted for in our analysis.

On the other hand, assume that the existing capacity will be exceeded at some future date and that a new increment will be built at that time, as shown in Figure 7-7. The new increment to capacity will now be required in year 3 rather than year 5. The increment would have been required whether the new project was executed or not. Therefore, the true economic effect of the project is to cause an expenditure to be made sooner. The new project does not *cause* the expenditure.

If a DCF analysis of the supporting-facility expansion were made on the assumption that the proposed project would be executed, the cost of the new increment, x, would be entered as follows:

Figure 7-7. New increment to existing capacity

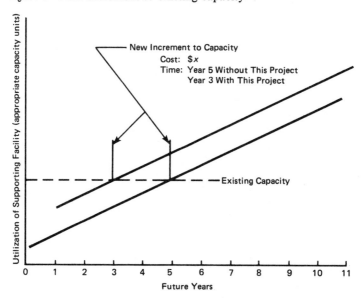

		With This Project				
Year:	0	1	2	3	4	5
Cost of new increment				(x)		

A companion DCF analysis made on the assumption that the proposed project would not be executed would appear as follows:

		Without This Project				
Year	0	1	2	3	4	5
Cost of new increment						(x)

Of course, this simplified case assumes no cost escalation between year 3 and year 5.

The phrase "charged to the project in the future years when additional capacity will be needed as a result of the decision to proceed with the project" requires explanation. There are only two circumstances that could require additional capacity: (1) Insufficient capacity presently exists; new capacity must therefore be provided immediately for this project. (2) Existing excess capacity will be insufficient in the future to meet the continued needs of this and other projects; new capacity must therefore be provided in a future year when it may otherwise have been unnecessary. It is obvious that the first circumstance is a special case of the second. The future begins at the moment the decision is made to proceed with the project. The question is: When will additional capacity be needed?

To answer this question, we must predict the needs of all other projects that would utilize a supporting facility. This is always difficult to do and is often impossible. The future competing needs for a utility system, boiler house, or general office building cannot be forecast exactly. In order to surmount this impasse, we can use a simplifying method of calculation that involves the concept of long-range incremental costs. These costs are approximations; since they are sometimes improperly used, it is important to clarify what they are approximations of.

If it were possible to predict the future utilization of a supporting facility, curves similar to those in Figures 7-6 and 7-7 could be drawn. The lines would not necessarily be straight, the lower line would not necessarily slope upward, and the two lines would not necessarily be parallel; but the curves illustrate the principles.

Ignoring depreciation for the sake of simplicity, we can compute the absolute differential for the project as follows:

With This Project Versus Without This Project

Year:	0	1	2	3	4	5
Cost of new increment				(x)		
10% discount factors	1	.909	.826	.751	.683	.621
Annual present value				$(.751x)$		.621x
Cumulative present value						$(.130x)$

NPV

Note that hastening the new increment has much less of an economic effect than charging the project with its pro rata share of the new increment. Suppose the proposed project utilizes a capacity equivalent to 75 percent of the new increment. To charge the project with this pro rata share in year 3, say, would result in an NPV of $-.751x$ $\times .751 = -.563x$.

The charging of a long-range incremental cost investment may be, and often is, seriously in error. But there is usually an offsetting circumstance—namely, the hastening of other new increments brought about by the proposed project for as long as the project continues to utilize the supporting facility. The situation is illustrated in Figure 7-8. If all three increments have the same size and cost, for example, the foregoing NPV reduces to $-.301x$. This NPV is beginning to approach the "pro rata" NPV of $-.563x$. If the project lasts long enough, the two may become nearly equal. This, then, is the justification for charging a new project with a long-range incremental investment.

In conclusion, be wary if your project can never afford the long-range supporting costs. Someone has to; you should too.

ESCALATION OF PLANT INVESTMENTS

Plant costs have risen sharply during various time periods, delaying projects and causing great management concern. Many factors have been responsible: inflation, other forms of escalation, problems encountered during the development phase of R&D, and increases in project scope.

Earl Oliver and James Moll have studied the capital cost increases for an oil shale plant using the TOSCO II process and a tar sands plant by Syncrude Canada Ltd. Their plot (typical of what projects encountered during the late 1960s and early 1970s) is given in Figure 7-9. The increases far exceeded the increases of the CE Plant Cost

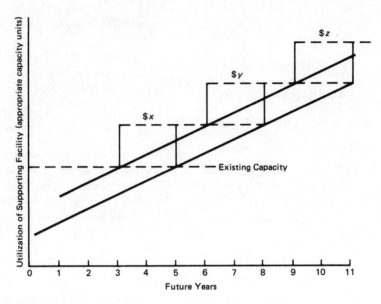

Figure 7-8. Serial increments to existing capacity

Figure 7-9. Capital cost increases

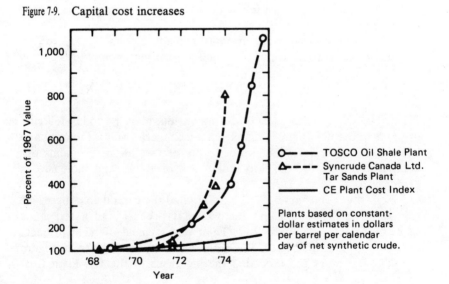

Index (published by *Chemical Engineering* magazine), which is intended to reflect the changes in cost of processing plants. The CE index correlates with a broader index of inflation, the GNP deflator. In fact, a plot of the GNP deflator would be indistinguishable from the plot of the CE index at the scale in Figure 7-9. How is the obvious total failure of cost indexes in this and other cases to be explained?

Oil shale and tar sands plants are special cases in that they use unproven processes, so we should inquire whether cost indexes have successfully reflected changes in costs of routine plants. The answer is a resounding no. For example, according to published studies, a 1,200-ton-per-day ammonia plant completed in 1967 cost W. R. Grace & Co. $33.6 million, compared with $107.4 million for a similar plant to be completed in 1978. The increase is 220 percent, far exceeding the probable increase of about 100 percent for the CE index. Ammonia production is well developed and relatively nonpolluting, so the increase should reflect escalation more than change in design (with the possible exception of investments needed to satisfy more stringent environmental legislation).

In 1974 plant costs and indexes diverged sharply: engineering contractors reported increases for petroleum and petrochemicals plants of 30 to 40 percent in six months, while the CE index rose about 14 percent. The 1974 experience is explained partly by the overheating of the economy at the time and partly by the derivation of the index. The CE Plant Cost Index depends on 67 Bureau of Labor Statistics (BLS) indexes, and all BLS equipment indexes depend on list prices. List prices are fictions that do not reflect contract prices in slack periods, when deep discounts are available. List prices are not likely to be raised as quickly as real prices when the economy improves.

Further, all indexes have one basic problem. They attempt the impossible in trying to assign a single representative cost to dissimilar plants when component costs are changing at greatly different rates—as happened between July 1973 and July 1975. The extraordinary increases for heat exchangers and centrifugal compressors reflect supply and demand. Many exchanger manufacturers left the business in earlier periods of low profitability, and many foundries shut down rather than comply with new antipollution and job safety rules.

Even the largest price increases do not explain the increases in cost estimates in boom periods because escalation clauses make final costs uncertain, schedule stretchouts increase costs of interest, insurance, and administration, and contingency allowances are likely to be increased. A related site-specific factor, local labor shortages, requires overtime pay and causes reduced productivity and further delays.

Our experience (and that of many others) is that investment cost estimates usually increase as processes advance from the laboratory stage to commercial use, even without the extraordinary factors of recent years. Costs may decline because of the discovery of a new catalyst, corrosion inhibitor, or the like; but optimism usually prevails until it is dispelled by hard data. After commercialization, unit costs often decline as larger plants are built and safety factors are reduced. This phenomenon is known as the *learning curve* and is shown in Figure 7-10. The hazard of comparing directly an estimate for an advanced concept with a corresponding estimate for an established process should be obvious.

The Boston Consulting Group has shown that for many products and industries amortized manufacturing costs excluding raw materials costs—also called value added—correlate well with cumulative production of an industry. Typically, this cost decreases 20 percent every time cumulative production doubles. In the early stages an industry frequently grows exponentially with time, so that a linear time scale may be superimposed on the logarithmic production scale. For capital-

Figure 7-10.   Learning curves

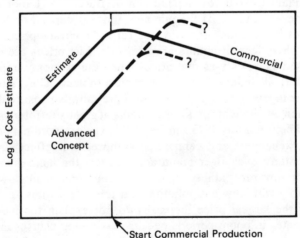

intensive industries, the capital cost tends to dominate the cost of value added, but raw materials are affected too much by extraneous factors to be correlatable.

Omissions and unresolved questions exist in all estimates. Table 7-2, which has been adapted from publications of SRI (formerly the Stanford Research Institute), may be helpful in searching for the factors that apply during various stages of a development program. The table does not quantify the cost uncertainties, but it can call attention to omissions and unresolved questions in an estimate.

How to account for the state of process development in an estimate is partly a matter of philosophy. On a statistical basis, less developed processes justify higher contingency allowances, but too much caution inhibits desirable research and development. The checklist in Table 7-2 should lead to better estimates in the early stages of a development program.

_____ WORKING CAPITAL

Working capital has for several years been regarded as synonymous with current assets. That is, it has been defined as the sum of cash, inventory, and accounts receivable. On occasion, it may include the counterbalancing effect of accounts payable and current liabilities. Whether or not payables are embraced in the definition of working capital, they spring from the same origins as the other three elements.

Perhaps the clearest way to define the elements of working capital is to describe how they come into being. In so doing, the important fact that all the elements are interrelated and interconvertible will emerge. At the same time, the groundwork will be laid for a discussion of ways to quantify working capital.

A corporation begins life with a supply of money borrowed from lending institutions and stockholders. At that moment, a balance sheet will show that the money reposes in an account called cash, on the left-hand side of the balance sheet, and is exactly balanced by the value of stock and debt on the opposite side. If the corporation is to become a manufacturer of commodities, one of the first things it will do is spend part of the cash for land, buildings, and equipment. On the balance sheet the money is transferred into the fixed-assets account, which is on the same side of the sheet and thus does not disturb the "balance."

As the fixed assets near completion, a supply of raw materials

TABLE 7-2. Estimates and scale of development

Scale of Development	Possible Unresolved Questions	Inaccuracies in Estimates
Bench, mockup, breadboard	All but main reaction or step	Only major unit may be considered. May neglect installation cost, engineering, and indirect costs; upstream, downstream, and storage costs; effect on existing equipment and capacity; cost of capital
Small pilot (process development)	Materials of construction Raw materials specifications Product specifications Catalyst life and cost Process materials life Effect of simulated feeds	May have major effect May require pretreating process May require purification process Sometimes major May affect operability as well as cost Trace components may require major process changes
	Effect of long-term recycle operations	May require major process changes or different materials of construction
	Effect of scaleup on yield structure	Affects capital as well as operating cost for a given product rate; scaleup of processes for solids more difficult than those for fluids

Large pilot, prototype, or demonstration	Environmental requirements	May have a major effect
	Complete integration with other processes	May create control problems
	Auxiliary-unit process costs	Historical data may not be valid because of changed requirements
Commercial	Changing ground rules: unavailability of raw materials, unavailability of fuels, new environmental regulations, political opposition	Delays from changing ground rules, strikes, and bad weather can be costly
	Remote area effects	Community and infrastructure development costs may be incurred
	Potential bias	Estimates from licensors and promoters tend to be low; those from companies seeking government support tend to be high
	Redundancy Capitalized development and operating costs Shortages and tight money	
	Technological advances	Estimates based on historical costs may be high

is purchased. The cost of the raw materials is placed in the inventory account, because payment rarely coincides with delivery. The balance sheet is brought up to date by placing the cost of the raw materials in a new account called accounts payable, on the opposite side of the balance sheet. As the payables are paid out of cash, the cash account and the payables both diminish equally, preserving balance.

In the meantime, operation of the facilities produces a supply of finished commodities, or goods. These goods come from the raw materials previously valued by the inventory account. That account is also redefined to include finished goods. However, the value of finished goods is greater than that of raw materials because of the labor expended in conversion. This additional value is matched either by a decrease in the cash account or by an increase in accounts payable. Either way, the balance remains.

When the finished goods are shipped to customers, the value plus the anticipated profit is transferred to accounts receivable, a new account on the left-hand side. The balance sheet is now out of balance by the amount of anticipated profit, and a counterbalancing account, called retained earnings, is placed on the right-hand side. (Retained earnings are derived from the statement of income and retained earnings.) As payment is received from customers, it is placed in the cash account, and the receivables account is diminished to the same extent. With this new supply of cash, the cycle is continued and repeated endlessly.

At any given moment after a business becomes established, a large number of separate cycles are in motion; they are never in complete synchronization. Balance sheets prepared at different moments will therefore show varying amounts of money in every account. These "snapshots" of the business at regular intervals make trends in the accounts visible, permitting appropriate actions to be taken.

The trends nearly always exhibit one significant fact: working capital as a whole is *growing*. As each new opportunity is acted on, additional money enters the working capital cycle, and the average value of each element of working capital grows larger. The analysis of a proposed project is therefore concerned with the incremental increases or decreases brought about by the opportunity.

Estimation of incremental changes in working capital elements should be based on a statistical approach that considers the dynamically interrelated nature of the elements. However, such an approach has not yet been developed. As a practical expedient, each working capital element is considered to be independent of the other elements. Further, the incremental change in the value of any component other than cash is assumed to be the absolute amount that the project would require if it were a self-sustaining business. Calculated values for inventory, payables, and receivables come closest to representing the effect of the project on the company. These elements are primarily under the control of separate company divisions; the total value of an element for the company is therefore the sum of the separate values.

On the other hand, even in decentralized companies, cash is primarily controlled by a headquarters group. At times, cash may be observed to fluctuate somewhat independently of the other elements. Corporate management must establish some minimum value for the cash needs of the company, and this minimum value must take cognizance of the money that all activities introduce into the working capital cycle.

Arbitrarily, therefore, the elements of working capital are defined and quantified as follows.

CASH Cash is defined as the average amount of money, in the form of bank deposits and marketable securities, that is necessary to support the operation of the company. Cash may be thought of as comprising three segments. The first is some minimum amount consisting of deposits in local banks and funds to cover casualty losses and other contingencies. New projects will have a relatively small effect on this segment of cash in an ongoing firm. The second segment represents funds available for long-term investment opportunities, income tax payments, and dividend payments. In effect, these are funds generated by existing projects and not yet disbursed. New projects will tend to increase this amount slightly.

The third segment is the most important from the point of view of a project analysis. It is the amount required to offset day-to-day variations in inventory, accounts payable, and accounts receivable. It is the "balancer" in the dynamic cycle of working capital. A project involving large purchases of raw materials ahead of need, or large cyclical buildups of product ahead of demand, will literally keep its cash working capital "working."

At times, cash will be visible as cash; at other times it will be tied up in goods. But it will be needed as long as the project exists; therefore, it will be unavailable for long-term investments. While it is temporarily visible as cash, however, it will be available for use as working capital in other projects whose cycles are not in phase with the project in question. As part of the same corporation, several projects may share the same cash, and the total requirement is not as large as it would be if the projects were independent.

If the inventory, accounts payable, and accounts receivable in a project were absolutely uniform from day to day, there would be essentially no requirement for cash working capital. Receivables collected produce cash, but most of it is used immediately for payables. Only the profit element of receivables raises the level of the cash account temporarily. The amount of cash included in an analysis depends on the cyclical demands of a project and on how accurately these demands can be portrayed for the other elements of working capital.

The cyclical nature of working capital cannot be portrayed with complete accuracy by either the annual-average method or the end-of-year method. In both the peaks and valleys, the day-to-day amounts of inventory, accounts payable, and accounts receivable are either obscured or overstated. Since DCF analyses are based on the calendar year, the peaks in the working capital cycle will coincide with the DCF interval only by chance. For example, if inventory has a peak in July, the end-of-year amount that would be shown on a balance sheet or an internal accounting document would be unrealistically low (Figure 7-11a).

If end-of-year inventory is employed in the analysis, a sizable amount of cash should also be introduced to cover the requirements for building inventory on the next cycle. However, the cash requirement would not be as great as the difference between the end-of-year inventory and the July peak. The reason is that other projects with countercyclical peaks would help to supply this cash. Similarly, if inventory has a peak in January, the end-of-year amount would be

Figure 7-11. Inventory peak in (a) January and (b) July

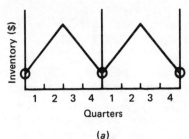

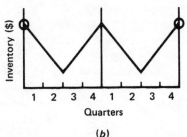

(a)   (b)

unrealistically high (Figure 7-11*b*). In this case, there should be a credit to cash, because the drawdown of inventory in the middle of the year will temporarily release funds for other projects.

The use of *average* inventory for cyclical cases such as these will compensate to some entent for peaks. The compensation will rarely be perfect, because the shape and duration of the peaks may not be adequately reflected by the average.

The inadequacy of present methods for estimating inventory and the counterbalancing cash requirements extends to accounts payable and accounts receivable. A perfect statement of cash would also allow for the cyclical natures of these accounts, as well as the countercyclical natures of all other projects. As a result, cash would be ignored— although obviously this cannot be so for large projects like the Alaska pipeline.

INVENTORY   Inventory is defined as the average value, at cost, of the combined supplies, purchased raw materials, product in process, and finished goods on hand that are necessary to support the operation of a business. As such, it represents the average amount of cash committed to maintaining sufficient stocks of these commodities.

There are two ways to evaluate inventory: LIFO (last in, first out) and FIFO (first in, first out). Each has a different effect on the income taxes paid by a business. Consider a company with an inventory of unused raw material of 1,000 pounds, valued at $1,000 on January 1. In January it buys an additional 1,000 pounds; but because of inflation, the price and value have doubled. In the course of the year it uses 1,000 pounds of the raw material, so that by the end of the year the inventory is back to 1,000 pounds.

Inventory is valued at $1,000 at the beginning of the year and increases to $3,000 during the year. Under the LIFO method, inventory returns to $1,000 by the end of the year (since the last inventory accumulated is assumed to be used first). Thus there is no effect on income taxes. Under the FIFO method, however, end-of-year inventory is valued at $2,000 (since the initial stock is assumed to be used first). The gain of $1,000 over beginning-year inventory is considered a profit and will cost the company about $500 in income taxes.

Despite the additional taxes, most companies prefer the FIFO method. The reason is chiefly cosmetic. Some company managements are willing to pay more taxes in order to show more profits to stockholders and the investing public. Apparently, even sophisticated investors—including large institutions (banks, insurance companies, mutual funds)—prefer to see company statistics that show more profits than are really there. They must feel that if a company uses realistic numbers, showing lower profits (on paper), the demand for and price of that company's stock will fall.

As a matter of custom, as well as law, the costs that go into inventory are those that visibly contribute to the value of materials in stock. That is, the costs involved in manufacturing and in moving a product between producing locations must clearly be seen to add to the value. When the product is placed in containers for shipment to customers, the cost of the container (if nonreturnable) is included in inventory. However, the cost of filling the container may not be included. Neither may freight costs to deliver the product to warehouses and customers and overhead costs such as advertising, research, and company administration. These latter costs are thought of as subtracting from income rather than adding to product value.

Ideally, inventory is computed as the average number of units of volume on hand over a given period multiplied by the average unit cost during that period. The period chosen for computing volume average may vary. Practically, it is often impossible to determine the volume necessary to support the operation of a business. Whereas the volume of finished goods can be estimated closely, it is more difficult, if not impossible, to estimate the volume of supplies (say, caustic soda) and product in process (the volume held on the trays of distillation columns).

Approximations are nearly always required. In general, the best methods of approximating the components of inventory are:

○ *Supplies.* Relatively small; say 2 percent of investment.

○ *Purchased raw materials and chemicals.* If purchased exclusively for the project being analyzed, estimate the number of weeks' supply that must be carried in stock. Two weeks is a point to start with; delivery schedules can have an important bearing on time. Value the materials at delivered cost plus unloading cost. If materials are purchased for several uses jointly, determine the ratio of the average value carried in stock to the total annual demand. Historical data may be of great help here.

○ *Captive raw materials.* If produced exclusively for the project being analyzed, estimate the number of weeks' supply that must be carried in stock. One week is a point of departure. Value at transfer price (often the market price) plus interlocation freights and loading and unloading costs.

○ *Product in process.* Relatively small; ignore, but check.

○ *Finished goods.* Estimate the number of weeks' supply that must be carried in stock. Two weeks is a point of departure, but be sure to recognize the seasonal nature of products such as fertilizers, which may have as much as six months' average supply. Larger inventory stocks are also required for materials such as polymer resins, which have many different grades. Also, do not forget the volume that may be in transit or in warehouses. Value at full production cost plus nonreturnable container cost.

For very rough estimates, the total inventory for a project may be based on a percentage of the annual production cost. For a product shipped in bulk, a point of departure is 20 to 30 percent of the annual production cost of the finished product.

ACCOUNTS RECEIVABLE   Accounts receivable are estimated by multiplying annual sales by the expected ratio of accounts receivable to annual sales. For many chemical products, the current ratio is approximately 11 percent. This means that nearly six weeks elapse between shipment of product and receipt of payment. Care should be exercised when the product in question is to be sold under terms markedly different from those offered for the majority of products. Agricultural chemicals have a much higher receivables ratio than do most chemicals. New products, in general, also tend to be higher.

Note that a project that has no sales has no receivables. Similarly, a comparison differential between two courses of action having the same amount of sales would eliminate receivables as well as sales from the comparison. Any pure cost reduction project such as a make-versus-buy opportunity has this characteristic.

In a one-year ROI analysis, the value of receivables is the average level achieved during that year. In DCF analyses and "average" ROI analyses derived therefrom, annual increments of receivables are shown as cash flows in the years they occur. For example:

Year	Sales	*DCF Schedule for Receivables Working Capital* (11% × Sales)
0	—	—
1	5,000	(550)
2	6,000	(110)
3	8,000	(220)
4	7,000	110

Year	Sales	DCF Schedule for Receivables Working Capital (11% × Sales)
5	8,000	(110)
6	8,000	—
		(880)

In a one-year ROI analysis, the value of inventory is the average level achieved during that year. In DCF analyses and "average" ROI analyses derived therefrom, annual increments of inventory are shown as cash flows in the years they occur. For example:

Year	Average Inventory Value During Year	DCF Schedule for Inventory Working Capital
0	—	—
1	500	(500)
2	600	(100)
3	800	(200)
4	700	100
5	800	(100)
6	800	—
		(800)

If the project concerns an existing business area, the existing level of inventory must be taken into account:

Year	Average Inventory Value During Year	DCF Schedule for Inventory Working Capital
Before Analysis	300	—
0	300	—
1	500	(200)
2	600	(100)
3	800	(200)
4	700	100
5	800	(100)
6	800	—
		(500)

ACCOUNTS PAYABLE   Accounts payable are an estimate of the average amount of cash owed for goods and services. As such, they represent cash that does not have to be removed from investable corporate funds. In the sense that a dollar not spent is a dollar earned, payables constitute a cash flow in a favorable direction and are therefore contrasted with cash, inventory, and receivables.

Working capital is *recovered* in DCF analyses by entering a working capital cash flow in the year following the last year of operation. This cash flow is equal to the ultimate level achieved during the operating life but is of *opposite sign*. In the foregoing examples, the recovery might be as follows:

New Business		Existing Business	
Working Capital Recovery in Year 7		*Working Capital Recovery in Year 7*	
Cash	720	Cash	420
Receivables	880	Receivables	550
	1,600		970

Note that inventory has been left out of calculations. This is a conservative assumption. Cash and receivables can generally be recovered. Inventory should be given little if any value.

—————————————— WORKING CAPITAL—FEDERAL TAX PRINCIPLES

There are two elections in the U.S. federal income tax law with respect to working capital:

○ Production costs are not deductible for tax purposes until the income from sale of the product becomes taxable. In the meantime, the costs are in inventory. There is one exception: inventory that becomes obsolete may be "written down" to whatever it is worth. The decrease in value is immediately deductible.

○ Income from the sale of a product becomes taxable at the time the invoice is rendered to the customer, which is practically coincident with the time of shipment. Between this time and the actual receipt of payment, the income is in accounts receivable.

An example will clarify how the tax implications of working capital are handled in DCF analyses. Although the example is simplified, it touches all the important points of a complex subject.

Assume that a manufacturing operation is set up to produce a limited run of a product whose only cost is the cost of raw materials. On July 1 of the first year of the analysis, the entire raw materials supply is purchased in one lot. The terms of payment are three months, and the cost is $2,000. On January 1 of the second year, production begins and continues at a uniform rate until the raw materials supply is exhausted on December 31. No product is sold in the year of production. On January 1 of the third year, shipment of the product begins and continues at a uniform rate until the supply of product is exhausted on December 31. The sales income from the product is $2,800 and the terms of payment are three months, with invoices rendered daily to the customer.

Figure 7-12 shows all the cash flows from this operation except for income taxes and cash working capital. Since shipment of the entire supply of product occurs in year 3, the entire $2,800 sales income is tax-liable in that year. Also, the $2,000 raw materials cost does not become deductible until year 3. Under this method of accounting, called *accrual,* the calculation of the actual tax (ignoring depreciation) is as follows:

	Year 3
Accrued sales income	2,800
Accrued cost	(2,000)
Actual taxable income	800
Actual tax: 52%	420

The schedule of the actual cash flow thus far quantified is:

	Year 1	Year 2	Year 3	Year 4
Sales income			2,100	700
Cost	(2,000)			
Tax			(420)	
Annual cash flow—less cash working capital	(2,000)	0	1,680	700

Working capital enters the picture because it is customary in DCF analyses to show sales incomes and costs on an accrual basis. That

Figure 7-12. Cash flows for limited run of product

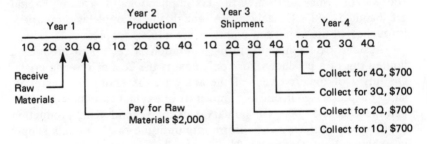

is, in order to avoid a separate tax calculation, DCF analysis does not usually show incomes and costs as they actually occur. (The only exception is stockpiling.)

A statement of actual cash flows for a project is, of course, much more difficult to construct than our simplified example, since production volumes, sales volumes, costs, and prices vary from year to year and since these variations occur concurrently. It is best to begin with a schedule of estimated sales volumes and assume that sales incomes, production costs, and so on, occur in concert. But this is the accrual method. And although the tax calculation will be essentially correct, some of the cash flows will be timed incorrectly.

Consider, then, the components of working capital in our example. Raw materials valued at $2,000 were received in one lot on July 1 and were untouched until January 1 of the following year. On January 1, the supply began to be used at a constant rate and was exhausted on December 31. The profile of raw materials inventory is shown in Figure 7-13a.

During year 2 product inventory was accumulated at a constant rate. During year 3 product inventory was depleted at a constant rate. The profile for the inventory of furnished goods (which is nearly always valued at cost) is shown in Figure 7-13b.

Accounts payable existed during the three months that elapsed between the receipt of raw materials and payment, as illustrated in Figure 7-14a. Accounts receivable first existed on January 1 of year 3, when the initial shipment of product was made. Each day of the first quarter, receivables increased. At the end of the first quarter, one-fourth of the product had been shipped. No payments had been

Figure 7-13. (a) Raw materials inventory. (b) Product inventory

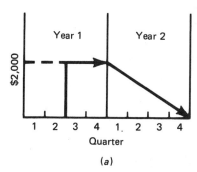
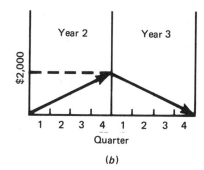

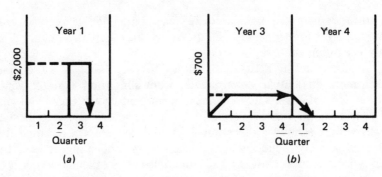

Figure 7-14.   (a) Accounts payable. (b) Accounts receivable

received, however, and the total receivables at that time was $700. Thereafter during year 3, collections for previous shipments matched new shipments, so that the level of receivables remained constant at $700 until the end of the year. All the available product was shipped by the end of year 3. Therefore, no new receivables were created in year 4, and the $700 level carried over from year 3 was brought to 0 by collections in the first quarter of year 4. The profile of accounts receivable is shown in Figure 7-14*b*.

Although it is not the customary practice in economic analyses, let us measure the components of working capital at the *end* of each year:

	Year 1	*Year 2*	*Year 3*	*Year 4*
Raw materials inventory	(2,000)			
Product inventory		(2,000)		
Accounts payable				
Accounts receivable			(700)	

The annual changes in working capital can be derived from the above data and set down as cash flows. To these are added the cash flows of sales income, costs, and taxes previously calculated on an accrual basis (see Table 7-3). This annual cash flow is exactly the same as was calculated on page 176 without resort to working capital!

A logical question arises immediately: Since working capital was introduced to simplify DCF calculations, and since the relatively simple end-of-year working capital produces the correct result, why is the more complicated average working capital employed? The answer is that the averages of the working capital components come closer to representing the cyclical cash demand than do the end-of-year

TABLE 7-3   Annual Changes in Working Capital

	Year 1	Year 2	Year 3	Year 4
Raw materials inventory	(2,000)	2,000		
Product inventory		(2,000)	2,000	
Accounts payable				
Accounts receivable			(700)	700
Total working capital	(2,000)	0	1,300	700
Sales income			2,800	
Cost			(2,000)	
Taxable income			800	
Tax: 52%			(420)	
Annual cash income			380	
Annual cash flow	(2,000)	0	1,680	700

amounts. Accounts payable in our example demonstrate the need for average working capital. Accounts payable did not exist at the end of year 1 and were therefore ignored. Furthermore, similar peaks could have occurred in accounts payable in other years or could have occurred in inventory and accounts receivable; they too would have been ignored. Those peaks would have required cash. Therefore, the practical expedient is to use the annual-average method, which does not completely ignore the peaks. A nominal and arbitrary amount of cash can then be entered, with the confidence that every reasonable effort has been made to reflect the economics of a project adequately.

_____ RAW MATERIALS

There are three price structures for raw materials: market price, alternate disposition, and replacement cost. The market price or market realization approach has already been discussed. Alternate disposition assumes the value of a diverted stream of new materials to be equal to its value in the product from which it was diverted less all costs associated with removing it from the product.

The replacement cost method of pricing is not limited to physical replacement (through purchase at some market price or through manufacture). Rather, it is a method of segregating the costs of products manufactured from a single raw material. Of course, there is no problem when only one product is involved, since the cost of the product is merely the cost of manufacture.

Replacement costing was developed in the petroleum, mining, and chemicals industries, where feedstocks and other raw materials are purchased and then processed into a large number of products. The feed cost and plan operating costs are a matter of record. But distributing these costs among the various products is a problem. Replacement costing is a way of realistically allocating the costs. Its advantage is that it segregates the cost of a single product without estimating the cost of all products.

The replacement cost method is based on two premises: (1) operations can be adjusted so that incremental raw materials yield only the primary product and one by-product; and (2) the incremental yield is a small percentage of the total primary product. By studying how costs change as additional raw materials are processed, the analyst can allocate the total costs to each product. Replacement costing reflects true costs only within the limits of a plant's operating range and is therefore idiosyncratic to each plant situation.

Replacement costing offers a good opportunity to illustrate the contrast between project economics and company economics. The differences between the two are almost always based on whether the feedstocks or raw materials are purchased at posted administrative prices or market prices. When feedstocks or raw materials are obtained from another company operation, they will already include some profit on the prior processing as well as on the primary raw material. Company economics should include the total company profit.

Most feedstocks and raw materials can be traced back to the point where they are in the ground. Therefore, when a company re-forms naphtha into ammonia, converts agricultural products to processed food, or recovers copper from an ore, two profits may be involved: the profit on the company's manufacturing operation and the profit, occurring elsewhere in the company, that is associated with producing the crude ore. Project economics would include only the first profit. Company economics would cover both. In effect we are saying that copper costs the company so many dollars per metric ton, and the cost to the company is less than this amount because of the production and transportation profits realized. We are thus looking at the overall company operation rather than just part of it.

When company economics are part of a project analysis, we often include them in the form of additional investments required for raw materials production and transportation facilities and the additional profit realized from raw materials production and the transportation. Then we add to the project economics the dollars-per-metric-ton profit to cover production and transportation. The result is the same as if we had developed the feed cost to the company.

_____ OPERATING COSTS

Normally, operating costs include all expenses incurred in the actual operation of a plant—that is, the costs of converting raw materials into salable products. Feed costs are not included in operating costs. Feed or raw materials costs plus operating costs are usually called manufacturing costs. But sometimes the terms "manufacturing costs" and "operating costs" are used almost interchangeably.

Operating costs for individual plants should be a matter of record for the company. Each plant should report annually at least a summary of its costs in a standard form, as shown below:

1. *Feed costs*

2. *Operating costs*
    Utilities and fuel
    Regular operating labor
    Administrative, clerical, and contract labor (if any)
    Maintenance labor
    Employee benefits
    Maintenance materials
    Contract maintenance (if any)
    Chemicals and supplies
    Plant services (labs, general facilities)
    Burdens (plant management, services, mechanical shops, materials procurement, technical assistance)
    Ad valorem taxes
    Depreciation

3. *Other Costs*
    Marketing and distribution
    Research
    Corporate overheads

Such a compilation will also indicate how important certain operating cost elements are in a business. For Du Pont, for example, selling, general, and administrative expenses (our "other costs") in 1975 amounted to 10 percent of sales revenue.

The total operating costs of existing plants should be available. But often we are concerned not with the operation of an entire plant, but with individual processing units or even individual pieces of equipment; or we are interested only in changes in operating costs with changes in levels of production. Therefore, it is necessary to break up costs and assign them to various operating units. Operating costs will then have to be built up from the components: manpower, chemicals consumed, utilities, and so forth. Each manufacturing plant has its own procedures for assigning costs.

As the items of expense included in total operating costs clearly show, many costs such as administrative expenses, technical expenses, and expenses for employee welfare cannot be assigned directly to individual operations. However, since the individual operations contribute to the requirements for these services, some system must be used for distributing the costs. Of necessity, the distribution will be somewhat arbitrary, but as accurate a method as possible should be devised. Remember, too, that any breakdown of costs between the various operating units is to some extent arbitrary. But the total money spent is real, and there is nothing arbitrary about the total.

The aim of project analysis is to predict operating costs for individual units or changes in operating costs with changes in levels of production. The best source of data for making these predictions is past experience on a given type of operation at a given location. To obtain total unit operating costs, we must add to those that can be directly attributable to the unit a somewhat arbitrary distribution of the costs that are not directly related to a unit's operation. This distribution should be as accurate as can be justified by the use that is to be made of it.

It is impossible here to undertake a detailed discussion of techniques involved in estimating operating costs. Even the literature provides only general guidance. Fortunately, remarkably good shortcut methods are available and suffice for most purposes. Studies made by SRI (formerly the Stanford Research Institute) indicate that for the chemicals industry annual operating costs (excluding raw materials, chemicals, and catalysts, but including depreciation) usually range within 30 to 40 percent of total investment. You can develop equivalent shortcut techniques by studying the available cost information for your particular company and industry.

Earlier, we introduced fixed (or period) and variable costs:

○ *Fixed (period) costs* are the costs associated with owning a unit versus the costs of operating it. Fixed costs have little relationship to the production rate. Depreciation, taxes, and insurance costs are good examples.

○ *Variable costs* change directly with level of production. If production doubles, these costs double. Chemicals, catalysts, and utilities usually fall into this category.

○ *Semivariable costs* fall in the middle ground between fixed and variable. They vary with the production rate, but not in direct proportion to it. Examples would include supervisors or staff and service people.

These classifications do not hold under all circumstances. Much depends on the particular situation. For a whole plant, operating labor is a semivariable cost. As throughput changes, the total number of employees will change, but not in direct proportion to the change in throughput. However, for a single operating unit or manufacturing station, the labor costs are almost fixed. Automation and continuous operation have tended to make manpower costs either fixed or semivariable rather than strictly variable, as they were historically.

Operating cost can also be broken down into direct and indirect costs:

○ *Direct costs* are directly attributed to an operating unit or manufacturing operation. They include costs of operators, maintenance, fuel, depreciation, chemicals, and utilities. Direct costs may be variable, semivariable, or fixed.

○ *Indirect costs* are the burdens and overheads associated with operating units or manufacturing operations. They are allocated on some reasonable basis; the specific method varies with different manufacturing complexes. Indirect costs are also called overhead or burden costs.

Total cost, of course, is just that: the total cost of operating a unit when all the direct, indirect, and fixed expenses are applied against the unit. It is the sum of all the items listed as manufacturing cost elements on page 181.

Sometimes it is desirable to know what portion of a company's operating costs is controllable (or changeable over a period of time) or what costs are associated with small changes in the operation of a plant. Controllable operating costs reflect the expenses of actually operating a unit but not the costs associated with owning it. The latter are the expenses that are incurred when the investment is made (regardless of whether the unit is operated or not) and that continue until the unit is retired from service. Normally, then, controllable costs include all costs except depreciation and taxes.

Burden costs are sometimes excluded in determinations of controllable costs. The assumption, of course, is that burden is a fixed cost independent of operations. For small changes in operations this may be true; but significant changes in operations usually have noticeable effects on burden costs, even though the effects may not be immediate. The problem here is mainly one of determining the direct relationship between operations and burden costs. Both of these definitions of controllable costs (including burdens or excluding burdens) are common in industry.

Incremental costs are those that reflect the effect on operating costs of small increases or decreases in the operation of a facility, such as changes in production rates, severity (quality control), and conversion. Incremental costs include only those that fall in the variable and semivariable classifications. They are peculiar to a given unit and the conditions under which it operates. For an existing facility, incremental operating cost is the difference in total operating costs before and after the change in operations. In the absence of actual operating data for a continuous manufacturing operation, we can assume as a first approximation that for a change in throughput operating labor costs are fixed, supplies and utilities are completely variable, and maintenance costs are 25 percent variable. Caution must be exercised in making any estimate of incremental operating costs; the figures are applicable only within the ground rules that were used to determine them.

Some points about operating costs should be emphasized. First, operating-cost estimating techniques differ widely in the manner in which burden or indirect costs are distributed over the operating units or cost centers. Uniformity in this area would be a great aid in cost comparisons and interplant studies within a company; but the diversity and complexity of a company's operations often make it impossible to lay down in a general accounting system hard-and-fast rules for collecting and distributing burden costs. Thus practices in this area vary somewhat from company to company and from plant to plant within a company.

Second, the method used in most companies assumes that the operating cost factors and correlations will remain essentially constant as new projects are added to existing facilities. This is a reasonable assumption in the light of past experience, which shows that these factors do not usually change significantly over short periods of time; but cost factors require watching in times of rapid inflation.

Finally, operating cost estimates are a prediction of average costs over a number of years rather than just the first or second year of operation. Ignoring inflation and assuming constant production rates, we would expect operating costs for a unit to be less during the early years of its life, since repair costs and efficiency-associated items are usually lower in a new unit.

DEPRECIATION

There is no single or simple definition of depreciation. A useful definition for our purposes relates 'to the corporate federal income tax. All business expenditures (except for purchases of land) are deductible from income in determining tax liability. However, some expenditures are deductible in the year incurred, while others must be deducted over a period of years. Expenditures for fixed assets fall into the latter category. The annual deduction from income of a portion of the cost of a fixed asset is called depreciation.

You will recall that the statement of income and retained earnings for Du Pont made no mention of the funds spent for fixed assets. In any given year, the expenditures for fixed assets do not cause any reduction in net income that year, except when depreciation has already started. In other words, stockholders have not yet "paid" for the fixed assets by having the cost of those assets deducted from sales income before net income is calculated and dividends declared. If the expenditures for fixed assets were deducted as incurred, it is quite conceivable that there would be no net income, or even losses, in some years.

Stretching out the deduction through depreciation is therefore a means of protecting stockholders from extreme variations in net income. Depreciation, although mandatory under most tax laws, would probably be voluntarily elected by most businesses. But note that depreciation represents a charge against sales revenue to recover money already spent. From a tax standpoint, depreciation is beneficial (compared with not being allowed to deduct any of a fixed asset's cost); the annual deduction is called a depreciation allowance. From the stockholder's point of view, depreciation is simply another cost of doing business; it is another deduction, another charge. (The law does not require that the schedule of depreciation allowances taken for tax purposes be the same as the schedule charged to stockholders. It is permissible for a corporation to keep two sets of books. However, for practical purposes, many firms use the tax-allowable rates for stockholder reports as well.)

There are numerous ways of calculating depreciation. The method or methods allowed for tax purposes are set by law and vary among countries and even among industries and individual projects in the same company. The method used for internal bookkeeping and reporting purposes is left pretty much up to the company, except for regulated industries, such as railroads and public utilities in the United States and the steel industry in the United Kingdom.

For internal accounting purposes, a company will set up a list of depreciation rates for use by all plants in a given country of operation. Companywide ground rules are established so that all plants use the given depreciation basis and rates. However, these rates are not necessarily the same as the rates used for tax purposes. Many countries allow higher depreciation rates or more rapid writeoffs as an incentive for new investment. And these are used for income tax reporting in order to minimize tax obligations.

Let us look briefly at three of the most common methods of calculating depreciation. These methods are specifically allowed by the U.S. Internal Revenue Code.

The first method is straight line, which is by far the simplest. Depreciation is charged in equal annual amounts to completely liquidate the original cost over the estimated life of the project. For instance, an asset costing $5,000 with a life of five years would be depreciated at $1,000 per year for five years, or 20 percent per year.

The second method is sum of the digits. Here depreciation is charged by the ratio of the years of useful life left in the asset to the sum of the numbers representing the total years of life of the asset. (For a five-year life, this sum would be $1 + 2 + 3 + 4 + 5 = 15$.) Thus in the example used above depreciation in the first year would be 5/15 of the $5,000 asset cost; the second year, 4/15; the third year 3/15; and so forth. This method assigns a greater depreciation to the early years of life, when the greatest decrease in value is likely to occur.

The third method is double declining balance (DDB). Here depreciation is allowed at twice the straight-line rate, but it can be applied only to the undepreciated value of the asset at the end of a fiscal period—that is, to declining balance. In the example above the depreciation rate, instead of being 20 percent, would be twice that, or 40 percent. Thus depreciation for the first year would be 40 percent of $5,000, or $2,000. For the second year it would be 40 percent of the undepreciated asset value—that is, 40 percent of $3,000, or $1,200. For the third year it would be 40 percent of $1,800, or $720. And so on. It is obvious that a 0 value can never be obtained by this method. To compensate for this, the IRS allows users of DDB to switch to straight-line depreciation any time they choose so that the asset can be completely depreciated. To get the maximum benefit of the accelerating features of the double-declining methods, users should make the switch when the depreciation charge under the straight-line method exceeds that possible using DDB. When the change is made, the straight-line rate is redetermined in order to depreciate the unrecovered cost of the asset equally over the remaining years of its useful life. In the example given in Table 7-4, the switch would be made in the fourth year.

Straight-line depreciation is used when percentage returns are calculated as average annual return on original investment (ROI). This is the only way that consistency can be maintained, since ROI gives a realistic picture of what is going on only when the annual savings are fairly uniform.

In discounted cash flow analyses of projects it is important to obtain the actual rates and the accounting method that will be used for depreciation for tax purposes. In most cases, the method will be the one that maximizes cash flows during the early years of a

TABLE 7-4. Methods of calculating depreciation

Year	Straight Line	Sum of the Digits		Double Declining Balance (DDB) = 40 percent of Declining Balance	DDB with Switch to Straight Line
1	$1,000	(5/15)	$1,667	$2,000	$2,000
2	1,000	(4/15)	1,333	1,200	1,200
3	1,000	(3/15)	1,000	720	720
4	1,000	(2/15)	667	432	540 Switch*
5	1,000	(1/15)	333	259	540
	$5,000		$5,000	$4,611	$5,000

*The DDB "book value" of the asset is $1,080 at the end of the third year.

project. Since depreciation laws are complex, the information needed will usually come directly from the plant involved.

It is desirable to know the effect that variations in depreciation method will have on the economics of a project. Different kinds of depreciation (straight line and sum of the digits), different depreciable lives (3 to 30 years), and different discount rates (5 to 20 percent) must be examined. To do this, we can take a fixed investment of $1 and calculate its total economic effect as measured by the cumulative present value. In all cases, the cumulative present value of the stream of investment and depreciation allowances will be negative. Obviously the less negative the better, since if there were no depreciation allowances the cumulative present value would be −$1.

One example is calculated in Table 7-5. The others are plotted in Figure 7-15. In order to simplify calculations, we have assumed that the investment is made in year 0 and depreciation schedules start in the year of investment. In all cases, the depreciation schedules are carried to completion; there is no lump-sum net book value or salvage value remaining.

A review of Figure 7-15 indicates that except for land, which does not depreciate, the economic effect of a fixed investment is always less than the face value of the investment. The initial investment is not recovered. Each annual depreciation allowance reduces the tax by 52 percent (48 percent federal tax and 4 percent state tax)—or by the effective company tax rate—of the depreciation allowance in that year. It is clear that straight-line depreciation is less attractive than sum-of-the-digits depreciation, and that raising the discount rate or lengthening the depreciable life reduces the beneficial tax effect of depreciation.

TABLE 7-5. Five-year straight-line depreciation of $1 investment

Year:	0	1	2	3	4
Depreciation	.2	.2	.2	.2	.2
Offset to other income before tax	(.2)	(.2)	(.2)	(.2)	(.2)
After-tax value of the offset (tax rate: 52%)	(.096)	(.096)	(.096)	(.096)	(.096)
Annual cash income	.104	.104	.104	.104	.104
Annual cash flow	(.896)	.104	.104	.104	.104
5% discount factors	1	.952	.907	.864	.823
Annual present value	(.896)	.099	.094	.09	.086
Cumulative present value	(.896)	(.797)	(.703)	(.613)	(.527)

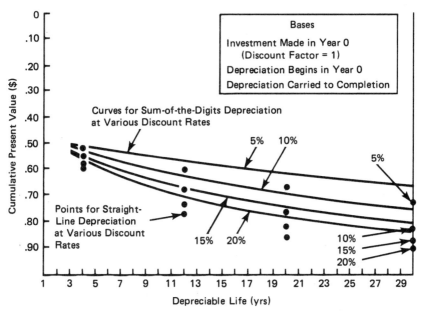

Figure 7-15. Depreciation of $1 fixed investment

Note that Figure 7-15 does not show depreciable lives shorter than three years. If the predictable life of an item is less than three years, IRS regulations generally permit the item to be expensed.

_____ INCOME TAXES

Complex tax technicalities make it difficult to give a single overall tax rate for all types of operating income. Many special tax factors, such as investment credits (special tax deductions intended to stimulate capital investment), accelerated depreciation, special reserve allowances, surcharges, excess profits taxes, and special exemptions, are applicable.

In the United States the normal corporate tax rate is 26 percent. In addition, there is a surtax of 22 percent on all profits over $25,000, making the maximum federal rate 48 percent for large corporations. (Some states also have small income taxes.) Formerly, a corporation that chose to consolidate operations of subsidiaries and affiliates for tax purposes had to pay an additional surtax of 2 percent on total corporate profits. This surtax is no longer assessed.

The income tax situation can get much more complicated in other countries. In addition to taxes on income, some countries levy taxes on dividends, interest, and profits transferred out of the country. Two tax rates are of interest in making economic studies:

The tax on company earnings paid to the local government. This rate is used in evaluating the effect of a project on the operating company's return.

Total taxes to the company. This is the total tax that the company pays and covers both the foreign tax and the U.S. tax on foreign earnings remitted to the U.S. The company receives a credit on its U.S. taxes for those taxes paid to foreign governments. This total tax to the company should be used to evaluate projects from a company's viewpoint.

The effective federal tax rate paid by a U.S. company can be substantially less than the 48 percent maximum, depending on the extent of foreign taxes paid or investment tax credits allowed. However, state and local taxes, generally amounting to an additional 4 percent, should also be allowed for.

The comparison of capitalizing versus expensing must be clear at this point. For example, if a $100,000 expenditure is expensed, it will result in a $52,000 tax saving (assuming a 52 percent tax rate) in the first year; if the same amount is capitalized over a ten-year period, the tax saving will be $5,200 per year for the next ten years. There is usually an incentive to expense when permissible, since cash savings made today are worth more than savings in the future.

chapter 8

Venture Analysis

A venture analysis for an R&D project must concern itself with the allocation of significant amounts of money and the plans for recouping that money, plus adequate profits from the cash flows generated during the economic life of the project. Decisions associated with the allocation of funds for a project normally are the responsibility of top management. These decisions are the major vehicle for implementing the strategic plan of the company. Once venture plans, including potential investment decisions, are committed, the decisions cannot be reversed without significant disruption to an organization. Venture decisions require careful analysis and are among the most critical and difficult decisions managers have to make.

A number of characteristics of venture analyses play an important role in evaluating the overall consequences of a decision.

Time. The commitment of large amounts of money is felt for many years in the organization. Thus the analysis of capital investment decisions should include long planning horizons (in some instances spanning 10 to 20 years) so that recognition is given to the time value of money (that is, a dollar spent or earned today is worth more than a dollar spent or earned in the future). Normally, the difference in value of money through time is taken into consideration by discounting the cash flows generated by the investment over the planning horizon. The discount rate used should reflect the opportunity cost that begins when the available funds are applied to the investment. When ample funds are available to an organization, the discount rate should be at least equal to the interest paid in borrowing money or to the return on investment from other ventures or programs.

Uncertainties. The long-lasting consequences of the investment decision create a problem of risks and uncertainties. Indeed, neither the cash flows that are expected over the life of the investment nor critical factors such as market shares and raw materials supplies are known for certain. Ignoring the risks and uncertainties is a severe oversimplification that may lead to selection of a poor investment alternative.

Risk. In an environment of uncertainty risk attitudes play an important role. It has long been recognized that investors are normally risk averse. The degree of risk aversion depends on their personal preferences, their present assets, and the nature of the uncertainties they have to face. A complete analysis of an investment must give decision makers the means to evaluate the implications of their risk attitudes in each strategy option being considered.

Tradeoffs among multiple objectives. In practice, the selection among competing alternatives often cannot be resolved by means of a single objective or attribute. In these cases, investment analyses must permit evaluation of the tradeoffs existing between multiple objectives.

Promoting and selling an idea requires a blend of quantitative estimates and recognition of the associated intangibles. As F. C. Jelen has observed, "These intangibles cannot be ignored merely because they cannot be assigned numerical values. The only certainty in a profitability study is that the profitability will evolve differently from that expected."*

Sophisticated managements recognize that the data supporting projections of sales, costs, earnings, and investments are never as

---

*F. C. Jelen, *Cost and Optimization Engineering* (New York: McGraw-Hill, 1970).

precise as the calculations. Typically, therefore, they will ask for alternative cases based on less likely but still possible circumstances. Sensitivity studies are one approach to allowing management to see the effects of selected deviations from base-case conditions. However, they do not provide a basis for estimating the likelihood that either the base case or the alternatives will occur. In addition, they do not portray the real-life situation in which all the variables are changing simultaneously at different rates and in different directions.

Tough as these problems are, they must be coped with. Rather sophisticated techniques of risk analysis have been developed that permit association of specific chances of attainment with any individual venture case. The fundamental notion of risk analysis, as we will see in Chapter 9, is that probability language can be used to describe risk and uncertainty (or its opposite, confidence). Mathematical techniques have been developed to allow subjective probability numbers to be manipulated in the same way as the more traditional or objective probabilities associated with dice and card games.

However, such analyses should not be undertaken lightly. Although it sounds sophisticated to be able to say, "This 10 percent DCF return project has a 98 percent probability of being better than 7 percent and a 40 percent chance of improving to a 13 percent return," it takes a lot of work and analysis for such a statement to have any meaning. Do not let the elegance of a technique deceive you into believing that the fundamental uncertainties have been surmounted. We suggest that you do not use these techniques until you're really ready for them. Using them too early will only lead to trouble.

Defining a good venture requires much creativity, not only to identify the most attractive opportunity for capital investment but also to construct all the relevant possibilities for approaching that opportunity. These efforts, which Herbert Simon refers to as the intelligence and design phases respectively, are perhaps the hardest parts of the decision-making process and the parts that have received the least attention in the literature. It is here that the natural talents of individual managers are tested the most.

Our approach provides a systematic way to structure the elements of a decision so that decision makers can make the best choice from the most promising alternatives. Another critical element in putting together a good venture analysis is identification of the decision makers, not only who the decision makers are in top management but also what they value relevant to the decision they will be asked to make or approve. This involves reaching some degree of agreement on the following:

The scope of the analysis.
The options that the decision makers want to consider.
Their attitude toward risks.
The tradeoffs they are willing to make at various levels of outcomes.

Agreement can be accomplished only through personal interaction with the decision makers, and the approach is significantly different with a single individual than with a group of individuals.

RELATION OF R&D EXPENDITURES TO CAPITAL INVESTMENT EXPENDITURES

Our decision-making criteria for R&D expenditures are an extension of the decision-making criteria for capital investment expenditures. Keep in mind, however, that there is an important difference (apart from the difference in tax treatments) between decisions involving investment in R&D, which purchases knowledge, and decisions involving investment in capital, which purchases plants and equipment. The difference is one of degree rather than of kind and resides in the greater degree of risk or uncertainty (or both) inherent in R&D decisions.

The terms "risk" and "uncertainty" are often used imprecisely as synonyms. Risk exists when a course of action (say, an investment decision) will lead to one of a set of possible outcomes (rates of return) with known probabilities. Uncertainty exists when the probabilities of the possible outcomes or the outcomes themselves are completely or partially unknown. The degree of risk can be measured by the dispersion of the probability distribution of the event whose value is being predicted. Uncertainty can be measured by the degree of lack of confidence that the estimated probability distribution is correct.

Capital investment is, in essence, present sacrifice for future benefit. Since the present is relatively well known, whereas the future is always unclear, capital investment also involves *certain* sacrifice for uncertain benefit. Expenditure for R&D can similarly be regarded as certain sacrifice for uncertain benefit. The greater uncertainty associated with R&D expenditure decisions is largely attributable to the fact that a longer projection into the future is required. The level of uncertainty attached to the estimates of sacrifice and benefit is accordingly higher.

For example, suppose a capital investment and an investment in R&D are made in the same year. The capital investment could involve expansion of an existing plant or a new plant capacity installation for an existing product. Or the investment could involve building a plant to manufacture a product new to the company, either through a licensed process or through the erection by a contractor of a "turnkey" plant. Another possibility is the erection of a plant to manufacture a new product created by past R&D activities. In all these possible forms of capital investment, the magnitude and timing of the "present sacrifice," if not certain, are fairly well known. The future benefits, while uncertain, are believed to commence in a reasonably short time.

The investment in the R&D project, on the other hand, purchases information that is intended to result in a capital investment. As the first step in the process leading to capital investment, the R&D project is automatically some years "behind" a capital investment project undertaken in the same year. Even after the technical feasibility of an approach is established, it may take five or more years of development engineering and related activities before capital investment is undertaken, as discussed in Chapter 7. The R&D project is therefore only the first link in a chain of expenditures of somewhat uncertain duration and amount. Further, the magnitude of the "secondary investment"—the capital investment required sometime in the future to implement the research results—introduces additional sources of risk and uncertainty.

More important than the greater uncertainty associated with the "purchase price" of R&D investment is the greater uncertainty associated with future benefits (assuming "equivalent product risk" with investment in capital assets. Again, this uncertainty arises from the fact that R&D spending will generally accrue over a more distant time horizon. The more distant a sales forecast, the higher the degree of risk and uncertainty associated with it. That is, the probability distribution of sales is wider; and the longer the forecast, the less confidence there will be that the estimated probability distribution is correct. This applies, of course, to problems in capital asset selection as well. Since ventures based on R&D activities *normally* involve longer time horizons than (similar) ventures not based on R&D, decisions involving the former will be characterized by greater risk and uncertainty.

Despite these differences, there is substantial similarity between R&D spending and capital investment spending, and the two compete for corporate funds. It is therefore appropriate to consider decision-making criteria for capital investment expenditures as a prelude to considering criteria for R&D expenditures.

_____ DECISION MAKING IN R&D PROJECTS

One of the most important problems facing a company is the selection of capital investment opportunities. The methods of selection ranged from highly subjective and intuitive approaches to more vigorous and objective approaches. Even the latter, however, require subjective judgment, since predictions of future occurrences are necessary. The difference is therefore one of degree rather than of kind. In the one instance, the judgments are translated directly into decisions; in the other, the insights are developed into explicit estimates of expected costs and benefits, which are then transformed into measures of profitability that form the basis for decision making.

The general trend in capital investment and R&D decision making is toward the objective approach. In essence, this involves the following steps:

○ Estimate the volume of sales, selling prices, costs of raw materials, operating expenses, capital investment requirements, working-capital requirements, selling expenses, and all other economic factors that have a bearing on the proposed project.

○ Summarize the basic estimates of project income and outlays in convenient form for appraisal purposes. (The computer is a useful tool here.)

○ Exercise managerial judgment in determining whether or not, in view of the projected business environment, (1) the anticipated return (expressed in whatever yardstick is chosen) is large enough to offset the business risks involved; (2) the investment opportunity is attractive in view of the various alternative opportunities for capital spending; and (3) the magnitude and timing of the investment are suited to the financial position of the firm and anticipated developments in the future.

The discounted cash flow (DCF) method, which we have stressed throughout this book, is concerned only with evaluating the basic estimates of annual incomes and outlays to arrive at a criterion for the worth of the project. There is nothing in the DCF method that makes it easier to prepare the needed estimates or to improve their accuracy. Similarly, nothing magical in the DCF technique relieves management of the need to exercise judgment on certain matters. The peripheral value of the DCF method here is that it focuses greater attention on estimates of the magnitude and particularly the timing of costs and benefits and that it offers a framework for better decision making.

A VENTURE EXAMPLE BRIEF AND SIMPLE

Remember that plastics resin you developed in Chapter 7? Let's make a first-pass venture analysis. Assume that the plastics resin is a relatively sophisticated one requiring uncommon and rather costly raw materials. Most likely, only narrow segments of the market will have a need for its unique properties. You do not see it competing directly with the billion-plus-pounds-per-year volumes of thermoplastic resins. Nor do you see it being moved at the low prices of these high-volume products.

From your knowledge of the competitive environment, you feel that a market of 50 million pounds per year could be created in a few years if the product were offered for sale at roughly $1 per pound. Using the techniques described in Chapter 7, you come up with a rough estimate of $50 million for plant investment. On the basis of that investment estimate, and the upper limits of the approximate total operating cost range, you estimate an operating cost of 45¢ per pound.

Obviously, at this point estimates of price, investment costs, and volume must be very crude approximations. Nevertheless, it is worthwhile to put these factors together into a DCF economic model and see what kind of story they tell. You want to find the answers to two questions:

○ Does the project have a chance to meet profitability targets, or should it be dropped now?

○ Is it worthwhile to go to the effort and expense to develop good estimates of the various economic components?

If you use the above data—combined with assumptions of a two-year construction period, a ten-year operating period, capacity production each year, an investment credit on taxes of 7 percent,

TABLE 8-1. Undiscounted cash flows for calculating sales price required to

Year	(1) Invest- ment	(2) 7% Invest- ment Credit	(3) 10% Depreci- ation	(4) Sales (million lb/yr)	(5) Sales (at $x/lb)	(6) Working Capital (25% of sales)
0	(25,000)	1,750				
1	(25,000)	1,750				
2			5,000	50,000	50,000x	(12,500)x
3			5,000	50,000	50,000x	
4			5,000	50,000	50,000x	
5			5,000	50,000	50,000x	
6			5,000	50,000	50,000x	
7			5,000	50,000	50,000x	
8			5,000	50,000	50,000x	
9			5,000	50,000	50,000x	
10			5,000	50,000	50,000x	
11[c]			5,000	50,000	50,000x	
11[d]						12,500x
	(50,000)	3,500	50,000	500,000	500,000x	0

[a]Col. 8 = cols. 5 + 7.   [c]Operating results.
[b]Col. 10 = cols. 1 + 2 + 3 + 6 + 9.   [d]Terminal value.

and a working capital requirement equivalent to 25 percent of sales—the DCF return would be 27 percent. Clearly, these are optimistic assumptions, but they are acceptable at this stage of your studies.

You should be cautiously encouraged by this finding. You can be sure that the rate of return will not be that high when you go forward with the project. But at least you have some room to maneuver. You have passed your first checkpoint. There appears to be sufficient margin over nominal minimum corporate DCF return criteria to allow for the future adversities you will encounter.

Knowing the inadequacies of your estimates at this point about price, costs, and investments, you use the same analytical framework to calculate DCF returns for a wide range of changes in these variables. Rather than calculating individual DCF returns, figure out and plot graphically the sales prices that would be required to give you 0, 10, 20, and 30 percent DCF returns. Do the same for the operating costs and the investments that would result in these levels of return.

Tables 8-1 and 8-2 show how to carry out these calculations to determine the price required to produce a 20 percent DCF return.

give 20 percent DCF Return (in thousands of dollars)

(7) Oper-ating Cost (at 45¢/lb)	(8) Profit Before Income Taxes[a]	(9) Profit After 50% Income Taxes	(10) Undiscounted Cash Flow[b]
			(23,250)
			(23,250)
(22,500)	$50,000x + (22,500)$	$25,000x + (11,250)$	$12,500x + (6,250)$
(22,500)	$50,000x + (22,500)$	$25,000x + (11,250)$	$25,000x + (6,250)$
(22,500)	$50,000x + (22,500)$	$25,000x + (11,250)$	$25,000x + (6,250)$
(22,500)	$50,000x + (22,500)$	$25,000x + (11,250)$	$25,000x + (6,250)$
(22,500)	$50,000x + (22,500)$	$25,000x + (11,250)$	$25,000x + (6,250)$
(22,500)	$50,000x + (22,500)$	$25,000x + (11,250)$	$25,000x + (6,250)$
(22,500)	$50,000x + (22,500)$	$25,000x + (11,250)$	$25,000x + (6,250)$
(22,500)	$50,000x + (22,500)$	$25,000x + (11,250)$	$25,000x + (6,250)$
(22,500)	$50,000x + (22,500)$	$25,000x + (11,250)$	$25,000x + (6,250)$
(22,500)	$50,000x + (22,500)$	$25,000x + (11,250)$	$25,000x + (6,250)$
			$12,500x$
(225,000)	$500,000x + (225,000)$	$250,000x + (112,500)$	$250,000x + (109,000)$

The unknown sales price is set at $x per pound. In summing up to get the undiscounted and discounted cash flows, make sure you maintain the integrity of your plus and minus signs. It is easy to make mistakes. Using parentheses to denote negative values is a good way to avoid confusion. By definition, the DCF rate of return is the percentage at which the present value of the cash outflow exactly balances that of the cash inflow; thus, if you set the sum of the discounted cash flow equal to 0, you can solve for your unknown quantity—the sales price required.

Similar calculations for 0, 10, and 30 percent returns generate the curve shown in Figure 8-1. The curve illustrates how sensitive your project return will be to different levels of price. Similar curves, calculated for changes in other significant variables, will help you understand the impact of conditions other than those you have built into your base economic model.

You have been looking, up to this point, at the best case possible: a plant that comes on stream at full capacity utilization and continues to operate that way throughout its useful life. This, of course, is

TABLE 8-2. Calculation of sales price required to give 20 percent DCF return (in thousands of dollars)

Year	Undiscounted Cash Flow	20% Present Value Factors	Discounted Cash Flow
0	(23,250)	1.	(23,250)
1	(23,250)	.833	(19,367)
2	$12,500x + (6,250)$	.694	$8,675x + (4,338)$
3	$25,000x + (6,250)$	.579	$14,475x + (3,619)$
4	$25,000x + (6,250)$	.482	$12,050x + (3,013)$
5	$25,000x + (6,250)$	.402	$10,050x + (2,513)$
6	$25,000x + (6,250)$	.335	$8,375x + (2,094)$
7	$25,000x + (6,250)$	.279	$6,975x + (1,744)$
8	$25,000x + (6,250)$	.233	$5,825x + (1,456)$
9	$25,000x + (6,250)$	.194	$4,850x + (1,213)$
10	$25,000x + (6,250)$	.162	$4,050x + (1,013)$
11^a	$25,000x + (6,250)$	.135	$3,375x + \quad (844)$
11^b	$12,500x$	.135	$1,688x$
	$250,000x + (109,000)$		$80,388x + (64,464)$

Sales price for 20% DCF return:
$$80,388x + (64,464) = 0$$
$$80,388x = 64,464$$
$$x = 80¢/lb$$

[a]Operating results.
[b]Terminal value.

not reality. In your particular case, involving a brand new product, you certainly are not going to jump right from the laboratory stage into a developed market filling up a 50-million-pound-per-year plant. It is going to take some time to reach ultimate capacity.

Let's look at two (of many) possible paths of growth into full capacity utilization and compare their DCF returns with the 27 percent return that results from full capacity operation. To do this, we need to build one more basic assumption into the model: that total operating costs are 50 percent fixed and 50 percent variable. Table 8-3 shows total sales at three different levels of capacity utilization.

The DCF returns for the 78 and 85 percent cases are compared with the full-scale output case in Figure 8-2. It is evident that the level of capacity utilization over the life of a project is critical to the attainment of satisfactory return. The price–cost–investment rela-

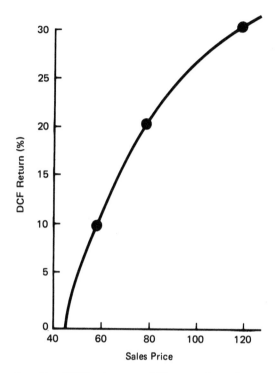

Figure 8-1.   DCF returns at different sales prices

tionships that give a 27 percent DCF return at full capacity lose about 5 of those DCF percentage points if sales average only 90 percent of capacity and lose much more than that as capacity utilization levels trend downward. The timing of the capacity levels is also critical.

SENSITIVITY ANALYSIS

The vulnerability of the DCF return to undercapacity utilization levels should make you wonder: "Is a 50-million-pound-per-year plant really the size I should consider building?" "Would I be better off looking at a smaller size so I could get up to capacity sooner?" "Is it more advantageous to build 30-million-pound-per-year plant that can be easily expanded to 50 million when the market is more certain?"

Don't guess at what the answers to these questions are. Calculate

TABLE 8-3. Sales at different growth paths to full capacity utilization (in millions of units)

	Sales Under Growth Path C		
Year	78%	85%	100%
1	10	10	50
2	15	25	50
3	25	40	50
4	40	50	50
5	50	50	50
6	50	50	50
7	50	50	50
8	50	50	50
9	50	50	50
10	50	50	50
	390	425	500

Figure 8-2. Project DCF return at different levels of capacity utilization

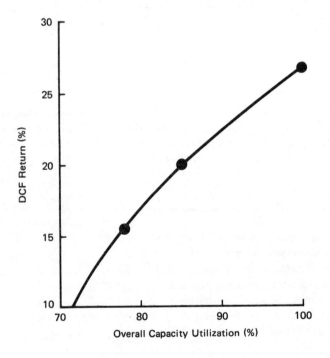

them. Prove to yourself that one case is better than the others you have under consideration. That's the beauty of discounted cash flow analysis. Because it brings into account the time value of money, it permits you to make choices among alternatives that span different actions taken at different times. A sensitivity analysis of the 30-million-pound-per-year plant is shown in Table 8-4.

Every concept you have can be tested; the testing is straightforward. Developing the concepts is critical. Study all the relevant alternatives you can think of before you decide on the best development plan.

CASH IMPAIRMENT PROFILE

Your project is going to put funds at risk. How much and when are the questions. For a 10-million-pound-per-year plant (the first unit you might build), you should develop a cash impairment profile, as shown in Table 8-5.

Remember the existing businesses of your company will have to generate the funds you need. Your $10 million initial investment will have to come from other operations' net of taxes (your investment

TABLE 8-4a. Base project data

Base Project Assumption	Assigned Value
Capacity (lb/yr)	30,000,000
Investment ($)	20,000,000
Sales price (¢/lb)	75
Maximum working capital ($)	4,000,000
DCF return (%)	16.3

TABLE 8-4b. Sensitivity analysis of data

Sensitivity	Impact on % DCF
10% investment change	±1.7
10% sales price change	±3.3
5% sales volume change	±1.3
10% working capital change	± .3
Lower fixed costs by $.5 million	± .4
Lower sales by 50% in first 2 years	−4.7
Assume 110% capability output attainable	+2.1

TABLE 8-5. Cash impairment profile ($000)

Year	Capital Investment	Investment Credit	Working Capital	Depreci- ation	Profit After Income Taxes	Cumula- tive Total
0	(5,000)	350				(4,650)
1	(5,000)	350				(9,300)
2			(1,000)	1,000	250	(9,050)
3			(500)	1,000	500	(8,050)
4				1,000	500	(6,550)
5			(1,000)	1,000	1,000	(5,550)
6				1,000	1,000	(3,550)
7				1,000	1,000	(1,550)
8				1,000	1,000	450
9				1,000	1,000	2,450
10				1,000	1,000	4,450
11[a]				1,000	1,000	6,450
11[b]			2,500			8,950
	(10,000)	700	0	10,000	8,250	

[a]Operating results.
[b]Terminal value.

tax credit of $700,000 will help), new borrowings or stock issues, postponements of other investments, or reductions in dividends. The "hole" dug by your project—the maximum negative cumulative total cash flow—should be a critical concern to all involved.

_____ WHERE DO WE GO?

Thus far you have probably developed enough information to discuss your situation with management. You and they know that every facet of your economic model will someday have to be studied in depth, and that the assumptions will have to be validated or replaced by more accurate estimates. The question at this moment is not "When do we get the money to build the plant?" Rather, the question is "Should we spend the money and the effort to go forward or drop it now?" You have done your homework properly. It was in your own self-interest to do so; after all, it was your brainchild that was under investigation.

Throughout the preceding chapters, we have emphasized the *planning* of a venture, largely from the perspective of the person who is responsible for developing an overall plan. The in-depth analysis of product function, markets, costs, and their uncertainties helped facilitate the choice between different development plans. The venture analyst—for that's what you are—was portrayed as a member of a development team.

In such situations, the reporting format is flexible and informal; proposals are made using the vocabulary, visual aids, and written style that the directly associated managements are comfortable with. Reporting is of a progressive character: each report starts where the previous one ended. Because of the gradual learning process, much of the report can focus on technical details of the venture structure.

At some point in the development of every new venture, however, a complete overview or *appraisal* is required to aid major decisions about the venture's future. Such appraisals are most frequently at the request of, or for, a decision maker who has not been intimately associated with the development. For example, an appraisal may be desirable to support a request for funds to build semiworks facilities. In this case, effective communication requires a sharper focus on the business elements of the venture than on the technical elements. In addition, a more standardized format may be necessary to permit comparisons with other ventures that management is familiar with.

Once you have done all this work, how do you best present it? Someday, it is hoped, you are going to write (with lots of help from a team) an appropriation request for a new facility. Start thinking about your project that way now. Summarize your studies (and the repeats you will make) in an appropriation request format like the following:

Body of the Report

Summary. Very brief presentation of the opportunity. Description of the project and/or the venture. Investment potential. Capacity potential. DCF returns. Management action request.

Background. Background information needed to understand developments. Why you regard this as an opportunity. The characteristics of the market you will serve, the competitive situation, prices.

Venture. What you have done thus far. Results achieved to date. How the project relates to previous concepts. Specific information about the project and its major underlying assumptions.

Relationship to Business Plan. How does the project relate to your company's corporate outlook and business plans? Is it reflected in those plans? Is it new? How does it compare with the next best alternative?

Timing. Why do it now? Schedule for the next phase.

Description of Proposed Work. Description of proposed work. Status of technology (including assessment of competitive technology position).

Manpower and Organization Implications. If any.

Legal Aspects. Potential patent position. Other aspects if pertinent.

Economics. Has DCF been calculated? Pricing assumptions, market projections, allowances for cost escalation, raw materials considerations, discussion of economic alternatives, sensitivities.

Risk Management and Environmental Considerations. Assessment of fire, employee injury, public liability, ecological factors, product risks, known safety problems, if any.

Supporting Data

Preliminary Appropriation Request Sheet. A one-page summary of salient information about the project, its investment, assumptions, economics.

Market Information Table(s). Total market projections. Assumed market share. Competitive projects' market shares.

Investment Table(s). Investment details.

Working Capital Requirements. Cash working capital. Raw materials, in-process and finished product inventories, accounts receivable.

Cash Profile. Project and venture details of net cash flow, profit–loss results, DCF returns, payouts.

Sensitivity Analysis. A table showing the impact on base-case DCF returns of changes in assumptions about major variables, taken one at a time.

Cash Impairment Profile. A table (or plot) of maximum amount company has at risk at any time.

chapter 9

Risk
and
Uncertainty

The discounted cash flow technique generates return figures that are consistent with those used in expressing bond yields. However, only for bond yields, where the magnitudes of the investment and the annual returns are known with certainty (excluding the possibility of default), is the DCF method rigorously applicable. That is, the method applies only when one possible value (and timing) can be assumed for each of the cash inflows and outflows, so that the probability of achieving the calculated return level is 100 percent.

This is clearly a considerable abstraction from reality as far as venture analysis is concerned, especially venture analysis for R&D projects. The estimates of the various cash inflows and outflows needed to implement the DCF method require forecasts of future events, which involve risks and uncertainties. The DCF method takes no cognizance of the existence of these risks and uncertainties, however; it simply provides a "one point" solution.

Attempts to incorporate risk into the DCF return or present worth criterion have resulted in considerable confusion. Most writers have been satisfied to treat risk intuitively or to pretend it does not exist. Only a few have really come to grips with the problem of defining and measuring it.

Among the unsatisfactory approaches used to allow for risk and uncertainty are shortening the expected life of the asset in the calculation; estimate earnings very conservatively; and using a higher discount rate for "riskier" projects. Another common technique, carried out in deference to the uncertainty inherent in capital investment analysis, is sensitivity analysis. As discussed earlier, this involves using the best estimates (or most probable values) of the various elements to determine profitability, and then calculating the effect on profitability of changes in the values. The utility of this method is limited, since it provides no guidance on the likelihood of occurrence of the various cases.

SUBJECTIVE PROBABILITIES

A much more elegant and useful approach to the treatment of uncertainty involves assigning subjective probabilities to the estimated values of the inputs. This in essence transforms the problem from one of decision making under uncertainty (or risk and uncertainty) to one of decision making under risk, a technique usually associated with Thomas Bayes and L. J. Savage. Subjectivists make the point that true or complete uncertainty is quite rare, and that it is possible to convert problems that seem to be under uncertainty (where there is no possibility of a probability distribution) into problems under risk. Subjectivists argue that probabilities are measures of subjective degrees of belief; hence the term "subjective probabilities." The estimates of such degrees of belief can, accordingly, be used as surrogates for probabilities when nothing else is available. Thus an event can be assigned a probability that reflects the decision maker's belief about the relative likelihood of the event's occurrence.

Typically, those who assign probabilities to the various elements of the profitability determination know a great deal about the possible levels, even when there is no basis for a frequency count. The subjective probability approach permits the utilization of this knowledge.

John Norton of Du Pont was an early proponent of subjective probability in evaluating new product ventures. He offers experimental evidence that subjective probability can be "amazingly accurate," especially when it is obtained from people who have had experience in similar projects. When personal experience is lacking, the subjective probabilities can be developed and improved through analysis of case histories and other analogies.

An early application of the subjective probability approach to the analysis of risk in investment decisions was made by Sidney Hess and Harry Quigley. They maintain that in making "best estimates" of the various elements entering a profitability calculation, the estimators are really implicitly thinking of a frequency distribution of the estimates. If the probability estimates are stated explicitly, a probability distribution of the profitability can be calculated from the probability distributions of the various elements. Hess and Quigley illustrate the approach with an example taken from the chemicals industry. A Monte Carlo simulation was used to carry out their calculations. In essence, this consists of determining some probabilistic property of a population of objects or events by taking a random sample of the components of the objects or events.

Among the elements considered by Hess and Quigley were plant capacity, fixed investment, fixed costs, marginal profit, sales volume, and selling price. Each element was represented by a probability distribution of values. A computer was used to draw a value at random from each probability distribution. This set of values was then used to compute a sample value of the profitability criterion, which in this case was net return on investment. Subsequent values of net return were obtained by additional random sampling of values of the elements. The process was repeated enough times so that the individual values of net return on investment approximated its "true" probability distribution.

IMPLICATIONS OF HIGH RISK FOR THE PROBABILISTIC APPROACH TO R&D DECISION MAKING

If estimates of the inputs to the profitability calculation can be made with a relatively high degree of confidence—that is, if the dispersions of the individual subjective probability distributions are narrow—a relatively narrow probability distribution of return levels around the most probable value will result. The decision on whether a project should be undertaken (or continued) could then be made fairly straightforwardly by comparing the most probable return level for the project with the company's target return level, taking into consideration the level of risk inherent in the dispersion of the probability distribution of returns. Similarly, it would be relatively straightforward to decide between the prospective return levels and associated risks inherent in alternative development strategies.

These concepts are illustrated by Figure 9-1a. The decision maker

Figure 9-1. Probability distribution of returns (target rate of return = 10 percent). (a) Project A: low rate of return. (b) Project B; high rate of return

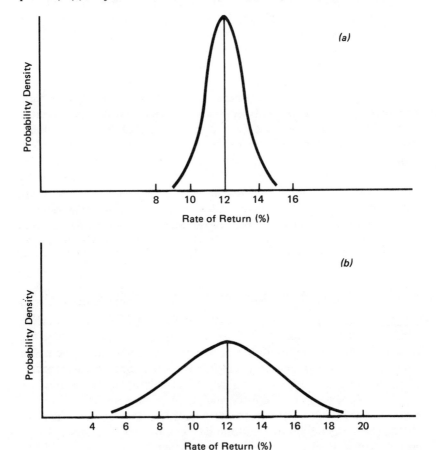

is considering Project A, whose probability distribution of returns is narrowly (and in this case symmetrically) dispersed around the most probable return level, 12 percent. Assume that the company's required return level is 10 percent. The decision to accept Project A is a "low-risk decision," in view of the fact that that there is a .95 probability (say) that the return will be greater than 10 percent. (A parallel argument could be made for rejection if the required rate of return were 14 percent.)

If the target return level is 12 percent, the problem is somewhat more complicated, but not terribly so. There is .5 probability of achieving this return level or better. The decision to accept the project may be made with the comforting knowledge that the probability of realizing less than a 10 percent return is only .05. Although the attitude of the decision maker toward risk—the aversion to a return of less than 12 percent—enters the decision, it does to a relatively minor extent with Project A.

In contrast, consider a subjective probability distribution in which some of the inputs to the profitability calculation are relatively broad; in addition, they include the inputs to which the return is particularly sensitive. Here, the calculated probability distribution of returns will have a relatively wide dispersion. Depending on the relationship between the most probable and the required rates of return, the decision to proceed with the project could be much less straightforward.

These concepts are illustrated by Figure 9-1*b*. The probability distribution of returns of Project B is also symmetrically dispersed around the most probable return level of 12 percent, and the required return level is again 10 percent. In this case, however, the distribution is quite broad. While there is still a .5 probability of achieving at least a 12 percent return, there is now a significant probability of .3 (say) that the return will drop below 10 percent, and a probability of .05 of realizing a 6 percent return on the project. Should the project be accepted if the required rate of return is 10 percent? This represents a "high-risk decision." In effect, the broad distribution of returns of Project B brings the most probable and the required rates relatively close together, whereas the narrow distribution of returns of Project A magnifies the differences between the two rates of return.

The matter of risk preference and risk aversion now comes very much into relevancy. Although there is a .3 probability of achieving a return of less than 10 percent, there is also a .3 probability of achieving a return greater than 14 percent, a .15 probability (say) of a return greater than 16 percent, and so forth. This quantitative specification of risk illustrates the power of the probabilistic approach. At the same time, it introduces an important issue that must be reckoned with if the approach is to be rationally implemented: What is the attitude of the decision maker (or of the company, if company attitude can be measured) toward risk? Does the prospect of achieving a 16 percent return, for example, outweigh (or balance) the equally likely prospect of achieving an 8 percent return? This question must be answered before the company can decide whether to accept Project B or whether, if Projects A and B are mutually exclusive, to accept Project A or B or be indifferent to the choice. (Although the returns in Figure 9-1 are symmetrically distributed, skewed distributions can occur if there is considerable interaction among the underlying alternatives. In such situations risk attitude becomes even more important.)

It is not our purpose here to discuss the importance of risk preference and aversion in decision making. Rather, our intent is to indicate that the probabilistic approach, for all its power in quantifying risk, still requires the decision maker's judgment as to the level of risk tolerable in light of the prospective return levels. In high-risk situations, the basis for making the decision may not be so clear-cut.

SOURCES OF RISK IN R&D DECISION MAKING

We must identify those elements entering into the financial analysis of a venture that contribute the most to the venture—that is, the elements that are most responsible for the dispersion in the subjective probability distribution of returns. These elements require special consideration in the estimating process. The major inputs (a more complete list is given in Chapter 5) to the financial analysis of a new product development project (the project of highest risk) can be classified as follows:

Engineering elements
Cost to develop a process
Time to develop a process

Manufacturing elements
Time required to design, construct, and shake down plant
Plant investment

Working capital requirements
Manufacturing cost

Marketing elements
Time and cost of market development
Marketing expenses
Product life
Sales volume as a function of time
Selling price as a function of time

Available evidence indicates that development projects are characterized by wide probability distributions of return. This is essentially the financial analyst's version of the common lament of the research director: that R&D activities involve high risk. Available evidence also indicates that the major source of risk resides in the marketing area. That is, the wide probability distribution of returns is usually due in large part to the inability to describe product sales volume and selling price within narrow limits. The high marketing risk can thus complicate the decision-making process, even if the subjective probability approach is invoked. In addition, the high marketing risk can obscure the effects of alternative development strategies on project profitability. (This problem is discussed in Chapter 10.)

SENSITIVITY OF PROFITABILITY TO THE MARKETING ELEMENTS OF A VENTURE

We can demonstrate the critical sensitivity of project profitability to the marketing elements—particularly sales volume and selling price—through some typical venture analysis calculations, letting estimates of the various elements vary about their "most probable values" and noting the resultant effects on calculated project profitability. Such calculations will usually show that small changes in price, sales volume, or distribution costs can have drastic effects on profitability, effects much more startling than large errors in final plant costs. Chaplin Tyler of Du Pont has developed a rough quantitative estimate of this sensitivity. He points out that overoptimism amounting to 10 percent in selling price and 10 percent in sales volume might readily bring about a 50 percent reduction in the estimated operating profit.

An analysis of three actual projects at PPG Industries, Inc. determined the sensitivity of project return to changes in four key factors: total investment, raw materials costs, anticipated sales price, and rate of sales buildup. Results were expressed as the percentage change in each of these variables required to cause a 1 percent change in the DCF return of the base case.

The results are summarized in Table 9-1. Project I was a large plant requiring an investment of $50 to $100 million for manufacture of a high-volume commodity. Project II was a moderate-sized plant requiring an investment of $2 to $3 million. Project III was a polymer facility requiring a $5 to $10 million investment.

It is clear from the PPG study that the project returns in these cases, covering a wide investment range, were far more sensitive to sales price and rate of sales buildup than to raw materials costs and investment. Changes of from 10 to 26 percent in raw materials costs and investment were required to effect a 1 percent change in the project DCF returns, whereas changes of only 3 to 9 percent in sales prices and sales buildup rates were sufficient to exert the same effect on profitability.

Further evidence of the critical role of marketing elements in chemicals product development is offered by George Schenk of General Electric. In examining the elements affecting the financial results of a new product venture, Schenk found that selling price most significantly affected the return on investment. Sales volume came next, followed by cost of sales and fixed capital investment. Schenk asserts that while there are individual sensitivities that will vary from project to project, the sequence holds true for a broad range of products.

TABLE 9-1. Sensitivity of DCF return to key project variables

| Variable | Percentage Change Required to Effect 1% Change in Project DCF Return | | |
	Project I	Project II	Project III
Sales price	3	3	6
Sales buildup (months to plant capacity)	6	5	9
Raw materials costs	12	13	22
Investment	26	10	15

RISK ASSOCIATED WITH ESTIMATION OF THE
MARKETING ELEMENTS

What are the sources of risk in engineering, manufacturing, and marketing?

Engineering risks arise in the technical development phase and are associated with the difficulty of translating known scientific principles into practical developments. They include the inability to design or develop the necessary manufacturing processes and the inability to correctly estimate the required time and/or cost necessary to develop the processes.

Manufacturing risks arise out of the difficulty of translating a working model into a full-scale production operation. This may occur because of the inability, within the time and cost limitations, to obtain necessary materials, equipment, technical personnel, or trained labor. Other manufacturing risks involve the inability to correctly estimate such elements as plant investment, working capital requirements, and manufacturing costs.

Marketing risks are associated with the difficulty of predicting the time and cost of market development, marketing expenses, product life, sales volume, and selling price.

Experienced R&D managers can generally predict fairly well the cost and time required for development. Even so, given the high sensitivity of project profitability to the marketing elements, marketing considerations are usually responsible for the major portion of the risk—that is, of the dispersion in the probability distribution of returns—characteristic of new product ventures.

A quantitative illustration of the degree of risk introduced by the marketing elements is provided by John Morovitz, Jr. Morovitz employed subjective probabilities to derive the probability distribution of returns for a potential investment project. The project was based on an R&D program in an advanced stage of development. The variables considered in the analysis were raw materials costs, raw materials utilization, plant investment, product prices, sales volume, and selling expense.

An engineering or marketing authority was asked to estimate a most probable value for each variable, as well as upper and lower limits, so as to embrace the range in which each variable could be expected to fall with .95 probability. Sound judgment based on the best information available at the time was used in establishing the confidence limits. The effects of these limits on the return on investment were then calculated. The frequency distribution of returns is shown in Figure 9-2.

The most probable return for the project was 9 percent, with a .4 probability of achieving at least this return. A return of 4 to 13 percent could be anticipated with a .95 probability. The dispersion was found to be attributable almost entirely to the marketing elements. To demonstrate this, Morovitz fixed the most probable values of the production variables at 100 percent probability and determined the variability of return attributable to the marketing elements. The resulting dispersion of returns around the most probable value was only insignificantly sharpened relative to the original case. There was still about a .4 probability of at least a 9 percent return and a .95 probability of a return within the 4 to 12 percent range.

Figure 9-2. Probability distribution of returns

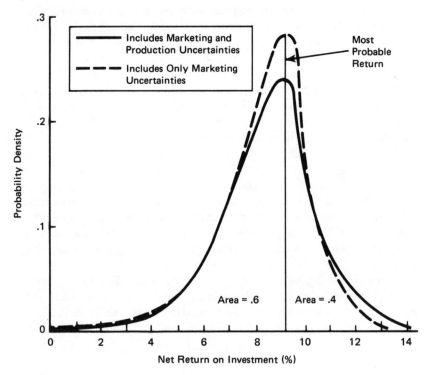

Even for a project in the advanced stages of technical development, then, the marketing risks far overshadowed the manufacturing risks. Had the analysis been carried out at an earlier stage of development, when the engineering risks would have come into consideration, there is good reason to suspect that they too would have been overshadowed by the marketing risks.

Further evidence of the relatively high level of risk associated with the marketing elements is to be found in an examination of the causes of new product failure (including failure to achieve predicted profitability). It seems reasonable to assume that the factors most responsible for new product failure are those that have a significant effect on profitability and those for which estimation is accompanied by a high level of risk. The evidence drawn from postinvestment appraisals, discussed in Chapter 2, clearly points to the marketing area.

Additional evidence can be drawn from the chemicals industry. Chaplin Tyler and Charles Winter of Du Pont analyzed 43 major investment projects (ranging from $1 million to $20 million, for a total of about $245 million) in an effort to determine why actual performance exceeded or failed to meet forecasts.

For 13 of the projects, earnings exceeded forecasts (by from 9 to 216 percent); higher than forecast sales volumes and selling prices were the principal reasons. For the remaining 30 projects, earnings failed to meet forecasts (by from 2 to 78 percent for 24 projects; the other 6 projects showed losses). As might be expected, lower than forecast sales volumes and selling prices were the principal factors in failure. For all 43 projects, marketing elements (sales volume and selling price) accounted for 35, or 80 percent, of all errors of estimate. In only 20 percent of the cases were the estimating errors attributable to nonmarketing factors (mainly faulty cost estimates and operating difficulties).

For the 30 projects that failed to meet forecast earnings, the primary reasons were:

	Number of Projects
Lower sales volume	16
Lower selling prices	8
Higher costs	4
Operating difficulties	2
	30

Some of these investment projects undoubtedly involved expansion of facilities for products already manufactured by Du Pont. But in light of Du Pont's new product activities at the time of the Tyler study, it is probably safe to assume that new products accounted for many of the projects considered. In any event, the marketing elements would introduce greater risk to the financial analysis of a new product venture than to the analysis of a product already manufactured by the company.

In summary, the profitabilities of new product development projects are seriously affected by the marketing elements, particularly product selling price and sales volume. These elements are the most difficult to estimate. As a consequence, we can concur with George Hegemen, who wrote:

> The main area of uncertainty in undertaking any new venture today involves the level of acceptance of a new product in the market. While our advanced engineering skills have largely eliminated the uncertainty in moving from the laboratory through the pilot plant to full-scale operation, I am sorry to say that our marketing skills have not advanced to the point where we are able to sharply reduce the uncertainties involved in introducing a new product to the trade. In fact, our research and development efforts, which create more and more new products each year, have only compounded the problem because the customer now has an increasing number of inter-changeable products to choose from. The effect of this has been to make the marketing of a new product more expensive and complex, and the risk and cost of failure is now higher every year.*

OPTIMUM PLANT CAPACITY

The concerns we have reviewed often lead to interest in determining the optimum plant capacity. When production output is forecast to be constant or growing slowly, the new capacity decision is not a difficult one. When a rapid growth in output is anticipated and there are significant economies of scale in the size of the plant, the problem of choosing the optimum initial excess capacity is more complicated.

Sound capital budgeting (assuming "perfect knowledge") would suggest that the cost-minimizing strategy is to make an investment when capacity equals demand, wait until demand has grown again

---

* "Marketing Models for Industrial Markets," paper presented at the Fifty-Eighth Annual Meeting of the American Institute of Chemical Engineers, Philadelphia, Pa., December 7, 1965.

to justify additional capacity, make another investment, and so on. The size of each investment and the time between investments are determined by two opposing factors:

Economies of scale provide an incentive to build the largest capacity feasible at a given time to reduce unit investment costs.

Cost of capital considerations provides an incentive to build the smallest plant possible at a given time.

Given the above considerations, the model developed by George d'Aspremont and his associates seems appropriate. This model can be regarded as an extension of studies and analyses of Alan Manne and Hollis Chenery. The model can be summarized by the following two equations:

$$x = \frac{2(1 - a)}{d(1 - a^2) + az_t}$$

$$\frac{c_{t+x}}{c_t} = e^d$$

where x = the time interval in years between successive investments
a = the capital scale factor in the relationships of investment as a function of capacity
d = the growth of demand
e = base for natural logarithms
z_t = the cost of capital at time t
c_t = the capacity of the plant built at time t
c_{t+x} = the capacity of the plant built at time $t + x$

The equations are plotted in Figure 9-3 for typical values of a ($a = .6$), d, and z_t.

Depending on the assumptions made about the cost of capital, the time interval between successive investments would be between two and four years at a growth rate (geometric) of 17 percent per year. However, this interval could be less than two years if the growth rate were perceived to be 40 percent per year. Capacity increments in all cases of interest would be expected to be twice the size of the previous capacity.

A number of limitations are built into the model (primarily to make the mathematics tractable). Two of the limitations are significant: (1) the model focuses solely on investments and ignores variable cost, and (2) no explicit allowance is made for technological improvements.

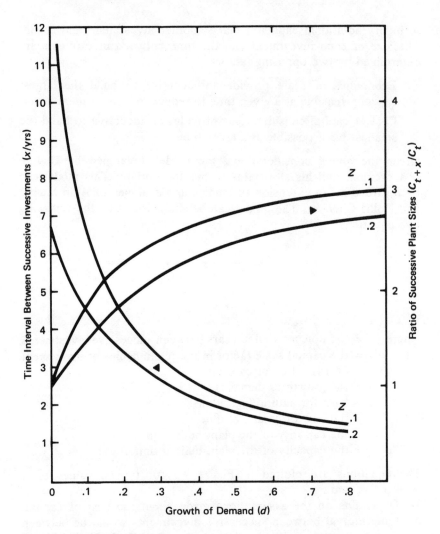

Figure 9-3. Optimum plant size

Nonetheless, even though their introduction into the model complicates matters, it does not change the answer significantly.

Once variable costs are introduced, the optimal capital budgeting policy does *not* necessarily involve a constant time between investments. Therefore, it is impossible to simplify the relevant equation, as was done in passing to the two equations above. Still, the equations can be used if *a* is regarded as the scale factor for total discounted cost. However, a would then vary depending on total production and therefore x_r would no longer be a constant. Simple manufacturing cost analysis expresses operating cost as a percentage of investment. Presumably, it is reasonable to assume a to be constant over some range of capacity.

Technological improvement may be viewed as changing the relationship between investment and capacity:

$$I_t = I_0^* c_t^a$$

where I_0^* is a constant (at a given time). If it is assumed that the factor I_0^* decreases at some rate, then

$$I_t^* = I_{0e}^{*-q(t-t_0)}$$

where q is the rate of technological change. The equations can still be used if z, the cost of capital, is replaced by $z + q$.

An analysis of pricing learning curves by the Boston Consulting Group shows that the major thermoplastics price is reduced by 20 to 30 percent each time cumulative production is doubled. This suggests a rate q of .05 or a total $z + q$ of .15 to .2. Considering all our problems and difficulties, this is insignificant compared with values of z of .1 to .15.

Error analysis shows that the optimum economic plant size is sensitive only to changes in market growth rate and the cost of capital. Errors in estimating plant capital costs and fixed costs have little effect on the point of inflection in the optimum capacity graph, even though the gross discounted costs do change significantly. In addition, the effect of discounting tends to negate errors, particularly in the later years of investment life.

For any particular cost of capital, the optimum plant investment cycle is a constant function of the production growth rate. For example, with a 12 percent after-tax cost of capital and a .6 scaling factor, an investment cycle of five years might be obtained. For a production growth rate of 10 tons per year, the optimum plant size would correspond to a capacity of 50 tons per year, thus preserving the integrity of the investment cycle.

The optimum investment cycle is not sensitive to small changes in the scaling factor, but it is particularly sensitive to changes in the cost of capital: the higher the cost of capital, the smaller the optimum investment cycle, and vice versa. These observations clearly identify the need to determine the cost of capital and to agree on production growth rates before making an economic analysis of capacity. In practice, production growth rates are likely to pose the greatest difficulty, since they are directly related to the uncertainties often associated with marketing new products.

QUANTIFYING RISK

The following principles are important in quantifying risk:

Each project variable typically has a range of possible outcomes. It is therefore helpful to develop, at least mentally, a probability distribution for each variable.

In most investment evaluations (except when risk analysis is employed) it is necessary to reduce the probability distributions to single "best estimate" values.

When a probability distribution is skewed, the analyst faces a choice between two common measures of central tendency: the model (most common) value and the mean (weighted average) value. ("Mean value" and "expected value" are used interchangeably to signify the weighted average of possible outcomes when the weights are probabilities.)

For example, in the following hypothetical distribution of sales volume, the model value is 5,000 tons per year and the mean value is 4,300 tons per year.

	Value (Probability)	Value (Probability)	Value (Probability)
Volume (tons per year)	3,000 (.4)	5,000 (.5)	6,000 (.1)

Since modal values are easier to visualize, there is a tendency to use them as "best estimates." However, *mean values are more appropriate in economic evaluations,* since they are mathematically unbiased estimates from the standpoint of single-point DCF calculations. The use of mean values tends to minimize the need for and the magnitude of risk premiums in hurdle rates (minimum desired rates of return).

Problems may arise when project variables are not independent of one another. For example, suppose the capacity of a plant is 100 units per period; there is an estimated .5 probability of 90 units per period of demand and a .5 probability of 110 units per period of demand. Expected demand (weighted average) is 100 units per period, but expected sales are only 95 units per period $[.5 \times 90] + [.5 \times 100]$ because of the capacity constraint. A return calculated with expected sales of 100 units per period would be overstated, since no inventory buildup was possible. Incidentally, in the above example it would be purely fortuitous if a capacity of 100 units per period were optimum with the given probability distribution of demand. A complete evaluation would include the investigation of alternate capacities with corresponding expected sales.

Simple interrelationships such as these can be handled in everyday project evaluations, provided careful attention is paid to the pertinent constraints. When the interrelationships are more complex, risk analysis may be helpful.

RISK ANALYSIS

Risk analysis builds on the basic concepts of cash flow analysis. A computer simulation model is used for a more explicit treatment of the elements of uncertainty. These elements are primarily related to the technological and commercial risks inherent in an investment. Inputs to the program are probability distributions for the project variables. Outputs, generated by a random-sampling technique, consist of a probability distribution of DCF returns and various statistics measuring the distribution's central tendency. Of these statistics, the most relevant measure for investment evaluation is the DCF return on the proposal's expected cash flow. An additional benefit of risk analysis, often overlooked, is that it requires a more thorough assessment of project risks in developing the probability distributions required as input.

Since risk analysis usually requires significant analytical effort and computer charges, it is not a universal tool. Its application is justified in projects when either the dispersion of possible returns is important or there are complex interactions between the project variables. Normally, the dispersion of possible returns given by risk analysis software is not important to the decision maker; the decision can be based simply on the DCF return for the expected cash flow. This statistic can be calculated without using the risk analysis software. Given the probability distribution for each project variable—and assuming that the interrelationships among the variables are not too complex—the DCF return on the expected cash flow can be computed quite accurately by using the mean of each probability distribution in a single-point DCF calculation.

The dispersion of possible project results is normally of importance to the decision maker only to the extent required to develop meaningful expected values. In some cases, however, some of the potential outcomes of a proposal might be considered unacceptable, even though these outcomes have very low probabilities of occurrence. Examples include actions that would exclude the company from participating in attractive future investment opportunities or contingencies that would threaten the company's financial stability or the interests of minority stockholders.

As with any other technique, the output of risk analysis is only as good as the quality of the input. Risk analysis cannot consider risks and uncertainties not explicitly reflected in the input data. The DCF return on expected cash flow generated by risk analysis will be just as biased as a single-point DCF if the same elements are ignored in both analyses. And the potential for being misled may be greater, because risk analysis looks and sounds very precise.

The DCF probability distribution generated by risk analysis is sometimes difficult to interpret, particularly when the rate of return is highly correlated with the value of a single variable and the dispersion of the variable is high. The result is that different sections of the DCF probability distribution are associated with quite different values for this variable. The danger signal is when the mean of the DCF probability distribution differs significantly from the DCF of the expected cash flow. The latter statistic is the better one for measuring the attractiveness of a proposal.

The assigned probability distribution of each significant project variable for a hypothetical venture proposal is presented in Table 9-2. Also shown for reference are modal and mean "best estimate" values for each project variable and single-point DCFs calculated using these values. Straight-line depreciation is assumed throughout, and there are no working capital requirements. The single-point DCF return using the modal values of the project variables is 11.5 percent.

After probability distributions are defined, relationships giving the net cash flow for each year of the project must be developed. In this example, the cash flow in year 0 will include only investment outlays; for all succeeding years the cash flow will be the same—the product of volume and price reduced by fixed and variable costs and taxes (after taking depreciation into account).

Since risk analysis is based on a random-sampling technique,

TABLE 9-2. Probability distributions of significant project variables

Variable	Value (Prob.)	Value (Prob.)	Value (Prob.)	Value (Prob.)	Value (Prob.)	Project Variable "Best Estimate"	
						Modal Value	Mean or Expected Value
Investment (million $)	8 (.15)	9 (.35)	10 (.25)	11 (.15)	11.5 (.1)	9	9.65
Volume (million lb/yr)	80 (.1)	90 (.2)	100 (.5)	110 (.2)		100	98
Selling price (¢/lb)	3.5 (.1)	3.75 (.15)	4 (.25)	4.25 (.3)	4.5 (.2)	4.25	4.09
Fixed cost (thousand $/yr)	400 (.2)	450 (.6)	500 (.2)			450	450
Variable cost (¢/lb)	1.6 (.2)	1.8 (.35)	2 (.2)	2.2 (.15)	2.4 (.1)	1.8	1.92
Tax rate (%)	50 (1)					50	50
Project life	15 (1)					15	15
Residual value in year 15	0 (1)					0	0

numbers from 0 through 99 are assigned to the alternative values of each variable in accordance with the assumed probability distribution. For the investment shown in Table 9-2, 15 percent of the numbers (0 to 14) are assigned to the $8 million value, 35 percent of the numbers (15 to 49) to $9 million, 25 percent (50 to 74) to $10 million, 15 percent (75 to 89) to $11 million, and 10 percent (90 to 99) to $11.5 million. Similar assignments are also made for the other variables. In this way the probability distributions of the input variables are incorporated in the numerous trials run by the risk analysis software.

At random for each trial, a number from 0 to 99 is selected to determine the investment level, another number from 0 to 99 to determine volume, another to determine selling price, and so forth. These values are then used to calculate the cash flow for the trial, from which the DCF return is determined. This process is repeated again and again. In addition, the cash flows of the trials are added algebraically and divided by the number of trials to yield the *expected cash flow*, from which the DCF return on the expected cash flow is calculated. In this example, 2,000 Monte Carlo trials were run with the computer.

The DCF return calculated for each trial is rounded to the nearest whole percentage and grouped to show the number of trials resulting in each different rate of return. The distribution for the example discussed here as well as the DCF return on the expected cash flow is shown in Figure 9-4.

Useful aggregate statistics can be developed from this DCF return distribution. Simple addition of distribution data, for example, shows a 50 percent chance that the project return will be in the 6 percent to 10 percent range and a 90 percent chance that the DCF returns will range between 3 percent and 13 percent.

The DCF return information can also be presented in cumulative form, as portrayed in Figure 9-5. This shows, for example, that 100 percent of the trials produced rates of return greater than −2 percent, about 50 percent of the trials yielded returns greater than 8 percent, and 12 percent yielded returns greater than 12 percent.

The DCF return on the expected cash flow of 8 percent is the relevant measure of central tendency for any evaluation of this project. The single-point DCF using the mean (expected) values of the project variables is 8 percent, the same as the DCF return on the expected cash flow generated by risk analysis. However, the single-point DCF using the modal values of the project variables is 11.5 percent. The 3.5 percent difference between the single-point DCFs arises from the skewness of the underlying probability distributions: upward

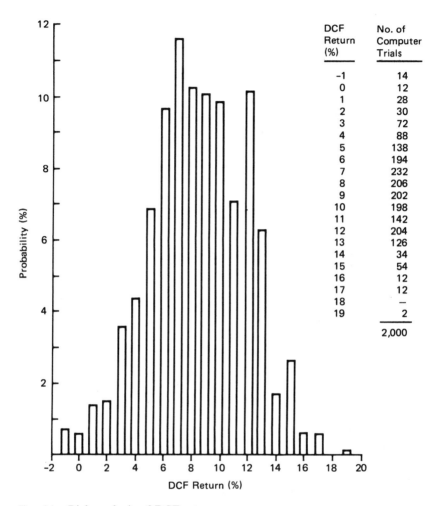

Figure 9-4. Risk analysis of DCF returns

skewness of the investment and variable-cost probability distributions, and downward skewness of the volume and selling price distributions.

In summary, when single-point DCFs are computed, the mean (expected) values of project variables, as opposed to the modal values, are the more appropriate "best estimates." When the dispersion of possible returns is not important to the decision maker, a single-point DCF using the mean (expected) values of project variables will usually be an adequate measure of attractiveness.

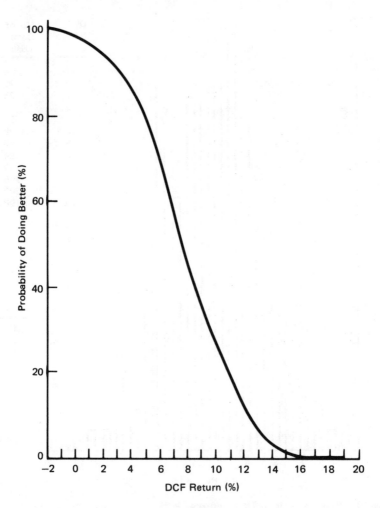

Figure 9-5. Cumulative DCF returns distribution

The example shown here has been deliberately oversimplified. In practice, risk analysis software contains many features designed to make the program realistic and useful. For example, we assumed no interdependence among the input variables, as mentioned above. A good program makes it possible to relate the values of two different variables, in effect making the first variable dependent on the second.

chapter 10

The Market Specification Approach

We have advocated a cautious probabilistic approach to R&D decision making, in line with recent developments in capital investment decision making under conditions of risk and uncertainty. As we have seen, however, venture analysis of R&D projects is significantly affected by the high level of risk introduced by the marketing elements. In this chapter we will develop a framework for venture analysis that focuses on the marketing area without using the probabilistic approach. A detailed venture problem is also worked out to illustrate the principles involved.

──────────────────────────────── A MARKET APPROACH TO RISK

The market specification approach (MSA) is founded on two principal considerations:

In the financial analysis of a potential venture based on a development project, the magnitude and timing of expenditures can generally be estimated with greater confidence than can the magnitude and timing of earnings.

Corporate managements usually invoke fairly definite financial yardsticks (such as specific rates of return) in making decisions on the allocation of funds.

The important feature of MSA is that, regardless of the method used to determine the target return level or the wisdom of the ultimate choice, there is little uncertainty associated with the target value, since the company itself sets it.

Conventional venture analysis treats project earnings (more precisely, product sales volume and selling price) as inputs to profitability calculation, with the profitability function, such as the DCF return, as the derived output, for subsequent comparison to the target level. MSA, in contrast, treats the company's target DCF rate of return and all the other elements entering the calculation as inputs to the problem; the project's earnings pattern becomes the variable to be determined. Thus, if the target DCF rate of return is set at the minimum acceptable level, and project expenditures are set at their expected values, the approach will specify the expected value of the *minimum project earnings—more specifically, the sales volume–selling price pattern that would have to be achieved in order to justify undertaking the project.*

The market specification approach, therefore, reframes the problem, substituting as an input a more certain quantity—a profitability criterion—for a less certain one—project earnings. By treating the major source of uncertainty as an output of the analysis, MSA brings an added dimension of judgment to the decision-making process.

The application of MSA to an R&D project aimed at a new product is as follows. When the project reaches a point where a business analysis of the total venture is needed, estimates are made of the size of the market for the new product and the expected penetration and selling price pattern. Also estimated are the cost to complete technical development, cost of market development, capital investment for a plant of the required capacity, working capital requirements, manufacturing costs, selling expenses, and so on. These estimates are used to calculate a most probable DCF rate of return for the project, for comparison with some target value, together with the probability distribution of returns.

Up to this point, we have followed the venture analysis approach for analyzing the financial outcome of an R&D project. In implementing MSA, we would *specify* the required DCF rate of return and use this target value, in conjunction with the expected values of project expenditures, to derive the marketing pattern that would have to be realized over the life of the project to fulfill the return criterion. The marketing pattern could be stated as the sales volume required for a projected selling price pattern, as the selling price required for a given projected sales pattern, or as the required yearly dollar sales volume.

Under the first alternative, for example, the output of MSA will be the specific annual sales volume required over the life of the project to provide the target return level. This volume will depend, of course, on the estimated project cash outflows, many of which (such as manufacturing costs, working capital requirements, and selling expenses) are themselves functions of the sales level. In addition, an infinite number of different sales curves (resulting in the same project present value) could be derived to satisfy the requirement; thus it becomes necessary to constrain the shape of the sales-versus-time relationship in some way to make a specific solution possible. The computation is somewhat complex and requires computer software.

It should be reemphasized that MSA is complementary to conventional venture analysis. In no way does it eliminate the need for ultimately estimating product sales volumes and selling prices through time. The expression of the results in terms of the *required* sales volume and selling price patterns, however, provides a tangible foundation for future market studies. MSA, therefore, focuses attention on the sources of greatest risk.

MSA provides a useful vehicle for judgment in deciding between alternate development strategies. The sensitivity of the venture to changes in the estimated expenditures and their timing is reflected in the corresponding changes in the required marketing pattern. Alternative development strategies can thus be viewed in terms of their respective required sales patterns (holding the selling price pattern constant).

We may wish to consider, for example, lengthening the development time and increasing development expenditures in order to improve some product feature. Here our intent is to broaden the market for the product at the originally forecast price. MSA quantifies the increment in sales volume necessary to justify the added expense and delay. The decision to undertake the proposed product improvement can then be clearly considered in light of the minimum marketing impact that the feature must produce. Marketing judgment can be used to assess the likelihood of achieving this incremental impact, or some multiple thereof.

_____ INPUTS FOR THE MARKET SPECIFICATION APPROACH

Application of MSA to a new product development project requires estimation of the following cash outflows and their timing:

Cost to complete development.
Market development expense.
Plant investment.
Working capital requirements.
Manufacturing costs (embracing raw materials and operating costs).
Selling expenses (including technical service activities).
Research requirements after commercialization (and/or royalty payments).
Any other expenditure incurred in implementing the venture.

The studies conducted for conventional profitability analysis should provide some idea of the attainable values of the marketing elements (product selling price and sales volume). However, estimates of the year-by-year working capital requirements, manufacturing costs, selling expenses, continuing research requirements during the life of product, and royalties (if applicable) are closely related to the level of sales (and, except for manufacturing costs, to the dollar volume of sales and thus to selling price). It is therefore necessary to express them as functions of sales volume and dollar volume, since sales volume and selling price are the variables to be determined by MSA. This implies that an iterative solution will be required.

SALES PATTERN There are, of course, an infinite number of sales patterns that could satisfy a return criterion. In order to make the problem determinate, we have to impose a constraint: an equation describing the shape of the sales curve.

The market demand for a new product characteristically grows rapidly at first and then tapers off as the market matures. This also applies to sales of a product that is new to the company but already mature, as the company approaches its equilibrium share of the total market. A number of models have been used to represent sales patterns exhibiting this gradually declining growth rate.

The model we use assumes that growth rate decays exponentially from an initial level to some long-run equilibrium rate, such as the growth rate for the national economy. The growth rate in year t is accordingly expressed by

$$r_t = r_\infty + (r_0 - r_\infty) e^{-kt} \tag{1}$$

where r_t = rate of sales growth in year t
r_0 = initial rate of sales growth
r_∞ = long-run (equilibrium) rate of sales growth
t = time of analysis (year)
k = rate constant (rate at which growth rate is declining from its initial to its equilibrium level)

The sales curve equation corresponding to this growth curve is

$$S_t = S_0 e^{r_\infty t + (r_0 - r_\infty)([1 - e^{-kt}]/k)} \tag{2}$$

where S_t = sales volume in year t (units)
S_0 = sales volume in initial year of analysis (units)

Good fits have been obtained in applying this model to a large number of industrial products. In Chapter 7 we applied such a model to low-density polyethylene. However, any sales model is acceptable.

In the implementation of MSA, product sales by the company can be represented by the Equation (2), with the shape of the curve dependent on the values of r_0, r_∞, and k, and the absolute sales volumes dependent on the level of S_0 as well. In one possible method of implementation, r_∞ would be set at some value representing the best estimate of the long-run growth rate of sales. Unless some reasons exist to the contrary, this value would normally be the long-run growth rate of GNP. A value for k would then be selected on the basis of historical experience with related products. For many products, for example, the rate of decline in growth rate is about 18 percent per year. Much higher rates of decline are possible, depending on the competitiveness of the market and the rate of technological innovation. Experience should provide a good basis for estimating the value of this rate constant for different product classes; it is a better basis, in any event, than exists for estimating the absolute levels of sales volume.

Finally, a value for S_0, the sales volume in the initial year of commercialization, would be estimated. Of all the yearly sales estimates, this is the least uncertain, since it occurs closest to the time of estimation. In addition, sales of an industrial product in the initial year are often determined not so much by the "inherent demand" for the product as by the scope of the market development program that is carried out. Market development will usually focus on a relatively small group of selected customers, who will provide the bulk of the orders during the first year. Thus in many cases first-year sales will depend on the resources committed by a company to generate sales during market development. Historical information on the first-year sales of related products and their corresponding market development programs should provide a good basis for selecting a value for S_0.

With values for r_∞, k, and S_0 set, only r_0, the initial rate of growth of sales, remains to be fixed in order that the sales curve be fully specified by Equation (2). This represents a logical unknown to be solved for in MSA. An alternative approach, of course, is to choose a value for r_0 on the basis of past experience and make k the variable to be solved for. The choice depends on which variable can be estimated more accurately for the product in question.

One further constraint on the sales curve must be set: sales cannot exceed plant capacity. This is expressed as follows:

$$(S_t)_{max} = C \tag{3}$$

where C = annual plant capacity

SELLING PRICE   New products fall into two general categories: those that are minor modifications of, identical with, or equivalent to existing products (for example, various polyester and acrylic fibers) and those that are entirely new products to the industry.

A company generally has much less freedom of action in pricing products in the first group than those in the second. When a closely equivalent product is available to meet the specific market need, the new product must be fit into the existing and anticipated future price structure, with due allowance for differences in quality, processing problems, and savings made possible by its use. In other words, the product may be new to the company's line but not to the market. The primary factor to be considered in pricing such a new product is the price of competing or substitute products.

Cost is far less important to pricing decisions than the above factors. Not only is cost largely unrelated to marketing considerations, but the true "cost" for a new product is extremely difficult to determine. Costs are so closely related to sales volume and to decisions concerning the allocation of research costs and startup expenses that much latitude exists in selecting a cost figure. Costs can play a role in the decision to manufacture the product, because they indicate whether the venture can be profitable in relation to the prevailing price structure; however, costs should neither determine nor influence the price.

In MSA selling price is estimated year by year by projecting marketing and competitive factors in the future. In this way, MSA resembles conventional analysis. However, it is possible to allow for decreases in selling price as sales volume exceeds appropriate levels. It is well known that an increasing physical volume of production is often accompanied by lower prices, at least in the early years of a product's life cycle.

Cost reductions can be accommodated in MSA by programmed price reductions or increases going into effect automatically in the year for which the Sales Curve Equation results in sales volumes exceeding appropriate levels. Allowance can therefore be made for "learning" as proposed by the Boston Consulting Group.

The output of MSA is the sales volume schedule necessary to justify the project at the required return level *for the selling price schedule in question.* MSA does not *predict* a selling price schedule; rather, it *derives* the sales volume–selling price relationship consistent with the required return. Marketing judgment can then be used to decide on the reasonableness of the required sales volume–selling price relationship and the likelihood of getting it under different development strategies.

MANUFACTURING COST   As discussed in Chapter 7, manufacturing costs can be classified into two categories: fixed and variable. Fixed costs are largely independent of the level of production (and hence sales) on a per-unit basis. In the case of a chemical product, they would include such items as raw materials, catalysts and chemicals, utilities, and *ad valorem* taxes. Variable costs include such operating costs as labor maintenance, plant overheads, insurance, and administrative expenses. Since they are essentially constant regardless of the level of production, they decrease on a per-unit basis as the level of production increases. Depreciation, research expenses, and selling expenses are handled independently in MSA and are not included as manufacturing costs.

Manufacturing cost per unit of production can thus be approximated as a function of capacity utilization by the following expression:

$$MC_f = k_1 + \frac{k_2}{f} \tag{4}$$

where MC_f = unit manufacturing cost at fractional capacity utilization f

k_1 = constant (fixed costs per unit of production)

k_2 = constant (variable costs per unit of production at full capacity operation)

f = fraction of capacity utilization (S_t / C)

If costs were truly either fixed or variable, Equation (4) would apply precisely over the whole range of capacities. In practice, this is rarely the case. A plant may be able to purchase raw materials, catalysts and chemicals, and utilities on more favorable terms when it is operating at 80 percent of capacity than when it is operating, say, at 20 percent of capacity. Generally, there is a critical fraction of capacity, where the relationship between cost and production rate changes. Thus these "fixed" costs per unit could display a slight variable element.

More significantly, labor (and maintenance) costs are not completely independent of the level of capacity utilization. The justification for assuming labor costs to be independent is that a labor force of fixed size is usually required to operate a highly mechanized or continuous-process plant, whether it is operating at 20 percent or 80 percent of capacity. But this is not always true. If the plant is part of a large manufacturing complex, for example, it could be run at 100 percent of capacity for 10 weeks during which sales are 20 percent of capacity. The plant could then be shut down, and the operators assigned to other duties for the rest of the year. However, in many plants union contracts, separation costs, and hiring costs sharply restrict the elasticity of the labor force. In general, labor costs should be considered "semifixed" costs.

In MSA this situation is handled by considering two regimes in the manufacturing cost–capacity relationship:

$$MC_f = k'_1 + \frac{k'_2}{S_t/C} \quad \text{for} \quad 0 \leq \frac{S_t}{C} \leq f_C \tag{5a}$$

$$MC_f = k''_2 + \frac{k''_2}{S_t/C} \quad \text{for} \quad f_C \leq \frac{S_t}{C} \leq 1 \tag{5b}$$

where f_C = critical fraction of capacity utilization.

For production rates below some critical fraction of capacity, labor (and other "semifixed") costs exhibit some elasticity and one set of constants (k_1', k_2') apply. When sales rise above this level, annual labor costs become fixed and another set of constants (k_1'', k_2'') become applicable. The constants are determined by the appropriate split between fixed and variable costs in the two regimes. Equations (5a) and (5b) thereby provide continuous expressions for manufacturing cost as a function of sales volume over the entire range from 0 to 100 percent of capacity operation.

WORKING CAPITAL Working capital depends on such factors as inventories (of raw materials, product, and operating and maintenance materials), accounts payable, accounts receivable, and cash working balance. Company standards may prescribe the guidelines for estimating working capital. For our present purposes, however, there is little point attempting a rigorous estimation. Instead, it is preferable to tie the working capital estimate to the dollar volume of sales.

Unless the project under analysis has unusual working capital requirements, therefore, MSA estimates working capital by allocating some percentage of the dollar volume of sales. The year-by-year working capital increments would equal that percentage of the annual increments in sales revenue, with the cumulative total being recovered in the final year of the project.

SELLING, ADMINISTRATIVE, AND RESEARCH EXPENSES Selling, administrative, and research expenses are the nonmanufacturing or overhead costs associated with the manufacture and marketing of a product. Selling expenses include the usual costs of marketing (salesmen, advertising, and so forth) as well as the costs of technical service in support of sales. Administrative costs are the share of central office management costs borne by the product. Research expenses may include the share of the costs of general research, unrelated to existing products, borne by the product, as well as the costs of the caretaker research needed to support the product after commercialization.

These costs can be expressed as a percentage of the dollar volume of sales. Each company will normally have its own guidelines, depending on the industry and the product line.

Our estimates should apply to the "steady state" years; during the first year or two, when sales are low, it would be meaningless to try to relate overhead costs to the sales level. It is preferable to estimate costs in the first few years independently, on the basis of the market introduction program or other company experience.

_____ AN EXAMPLE OF THE MARKET SPECIFICATION APPROACH

The product in question is assumed to be both new to the company and new to the industry. The research phase of the R&D project began in 1963 and continued at a low level through 1965. Through the end of that year, research costs amounted to $300,000, which was to represent about 10 percent of total R&D expenditures for the project through the time of commercialization.

Development commenced in 1966, which is considered the initial year of analysis. The product was by this time reasonably well defined, and the rudiments of a process had been demonstrated in the laboratory. In this same year, the decision was made to build a pilot plant and commit substantially greater resources to the development of the process and product. This is clearly the year in which a financial analysis of the project would be undertaken.

R&D costs from this point were as follows:

1966	$800,000
1967	700,000
1968	500,000
1969	400,000
1970	400,000
1971	400,000

Plant design studies were conducted in 1966–1967, and $3 million was appropriated in 1968 for the construction of a plant with a capacity of 50 million pounds per year. Construction began in 1968, and the plant was in service by 1970. Market development activities took place over this period. Some $600,000 (net) was spent on producing market development samples in the pilot plant facilities, and another $200,000 was spent on promoting and supporting customer evaluations of the product. An additional $400,000 in miscellaneous expenses (startup expenses, early selling expenses, and so forth) was incurred in support of the project. This $1.2 million total was incurred substantially uniformly over the years 1968–1970.

The plant is part of a large complex producing a wide range of products. The labor supply is relatively flexible. As a result, the manufacturing cost is relatively inelastic with respect to production rate. Out-of-pocket manufacturing cost (exclusive of depreciation, selling expense, and research charges) is 15¢ per pound at full capacity. At 25 percent of capacity (12.5 million pounds per year) manufacturing cost is 24¢ per pound, and at 10 percent it is 36¢ per pound. The manufacturing cost expressions used to approximate actual behavior are accordingly:

$$MC = 16 + \frac{2}{S_t/C} \text{ ¢/lb} \quad \text{for} \quad 0 \le \frac{S_t}{C} \le .25 \tag{6a}$$

$$MC = 12 + \frac{3}{S_t/C} \text{ ¢/lb} \quad \text{for} \quad .25 \le \frac{S_t}{C} \le 1 \tag{6b}$$

The relative inelasticity of manufacturing cost is indicated, for example, by the fact that doubling the output, from 50 to 100 percent of rated capacity, drops manufacturing cost by only 17 percent, from 18¢ to 15¢ per pound.

The introductory selling price of the product in 1970 was 30¢ per pound. The price was set on the basis of the market development program, which established a limit of what the market was willing to pay. By 1973 the price was dropped to 28¢ per pound. It is anticipated that by 1980 the price may drop to 26¢ per pound (considerations of inflation aside).

In 1970, the first year of commercialization, 1.5 million pounds of product was sold. The constant k, describing the rate at which sales growth rate is declining from its initial level to its equilibrium level (assumed to be 5 percent per year) is about .3. The sales volume equation employed in the base case (with r_0 left unspecified) is therefore:

$$S_t = 1.5 \, e^{.05t + (r_0 - .05)} ([1 - e^{-.3t}] / .3 \text{ million lb/yr}) \tag{7}$$

$$(S_t)_{max} = 50 \text{ million lb/yr} \tag{8}$$

The inputs to the base-case analysis are summarized in Table 10-1, which assumes that product sales will continue through 1984 (for a total product lifetime of 15 years). Utilizing MSA, we can calculate the annual sales volumes, consistent with sales volume restraints implied by the above equations, that will provide a DCF rate of return equal to the target level (assumed to be 10 percent for this illustration).

The solution is accomplished iteratively, with computer software written by Howard Oakley of Exxon Research & Engineering Co. The software is highly efficient and results in a rapid convergence

TABLE 10-1. Summary of inputs to base-case market specification analysis
(in thousands of dollars)

Year	R&D Expense	Invest-ment	Market Development and Miscellaneous Expenses	Selling Price (¢/lb)	Sales Volume (million lb/yr)
1966	800				
1967	700				
1968	500	3,000	400		
1969	400		400		
1970	400		400	30	1.5
1971	400			30 ⎫	
1972				30 ⎬	To be
1973–1979	5% of			28 ⎭	derived
1980–1984	sales			26	

Manufacturing cost:
$$MC = 16 + \frac{2}{S_t/C} \ \text{¢/lb} \qquad \text{for} \qquad 0 \le \frac{S_t}{C} \le .25$$

$$MC = 12 + \frac{3}{S_t/C} \ \text{¢/lb} \qquad \text{for} \qquad .25 \le \frac{S_t}{C} \le 1$$

Working capital factor: 25% of dollar volume of sales
Selling expense factor: 12% of dollar volume of sales
Research factor: 5% of dollar volume of sales
Sales volume: $S_t = 1.5 e^{.05 + (r_0 - .05)} [(1 - e^{-.3t})/.3 \ \text{million lb/yr}]$
$(S_t)_{max} = 50$ million lb/yr
Product lifetime: 15 years (1970–1984)
Depreciation: Sum-of-the-digits method, 11-year depreciable life
Investment credit: 7%
Salvage value: Equal to dismantling cost
Federal income tax rate: 50%

to the solution. Generally, solutions are obtained in less than five iterations, and the resulting sales curve is consistent with the desired DCF return level to within a tolerance of ± .2 percent (that is, 10 ± .2) of the specified level. The calculations are based on annual discounting of the end-of-year cash flows.

REQUIRED SALES VOLUME FOR BASE CASE The calculated sales volumes required to furnish a 10 percent discounted cash flow rate of return for the base case are shown in Table 10-2.

TABLE 10-2. Sales volumes required for 10% DCF rate of return,
base case (in millions of pounds per year)

Year from Commercialization	Required Sales
1 (1970)	1.5
2	3.9
3	7.9
4	13.6
5	20.5
6	28.3
7	36.3
8	44.3
9–15	50

In this case, the sales pattern developed by MSA for a 10 percent rate of return does not appear to offer an unreasonable target. The plant, which is sized at 50 million lb per year, need not achieve full capacity operation until year 9 of commercial operation. Many plants are sized to "fill up" after five years of operation. To achieve a 10 percent return, the plant in question need reach only 40 percent of capacity by year 5 (provided, of course, that the rest of the required pattern is fulfilled). The 10 percent return pattern is shown in Figure 10-1.

In practice, the company can consider adding to plant capacity in year 8. If a sequential investment is contemplated during the development process, at the time the initial analysis is being made, it can be easily accommodated by MSA. Alternatively, the approach can simply be invoked again to analyze the marketing implications of an additional investment.

A FRAMEWORK FOR CONSIDERING PROJECT RISK   In addition to determining the sales volumes required to achieve the target 10 percent rate of return, it is important to consider the required sales curves for rates of return close to the target level. The results can provide a useful framework for understanding project risk. Figure 10-2 shows the calculated sales curves consistent with rates of return of 8, 10, 12, and 15 percent for the base case.

To appreciate the utility of this portrayal, consider once again Figure 9-1 (page 210), which shows the probability distribution of returns for two projects. The wider distribution for Project B was

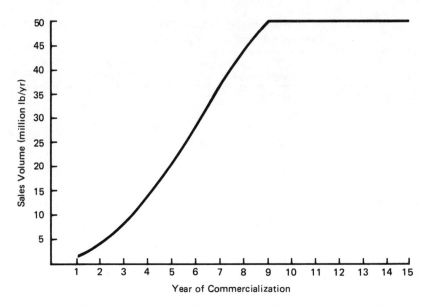

Figure 10-1.   Sales volumes required for 10 percent DCF rate of return, base case

more typical of the distributions facing the R&D decision maker. But the wide distribution is not enough to make this a "high-risk decision." The decision to undertake Project B was characterized in these terms because of the broad distribution *and* because of the relationship between the most probable and the required rates of return (12 and 10 percent respectively). The broad distribution of returns associated with the project (which is most commonly due principally to marketing uncertainties) tends to diminish the ability to discriminate between these two return levels.

Let us assume for the moment that Figure 10-2 refers to Project B. In effect, the figure expresses the difference between achieving the most probable return of 12 percent and achieving the target return of 10 percent in terms of the difference in the sales curves consistent with each of these return levels. The technique thus exposes the "latitude," and thereby the "risk," inherent in the project in terms of sales volume. This element, along with selling price, is the one most responsible for project risk. Since the selling price pattern is held constant in calculating both sales curves, this technique is a

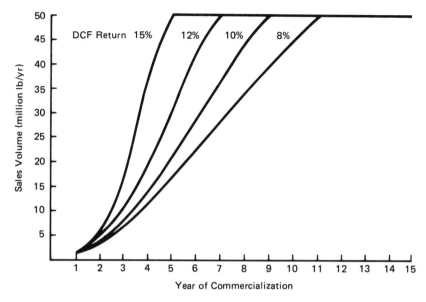

Figure 10-2. Sales volumes required for different DCF return levels, base case

most appropriate way to illuminate project risk.

For example, to achieve the company's target return level of 10 percent, sales must reach the rate of 50 percent of plant capacity after 5.5 years of operation and 100 percent of capacity after 9 years of commercialization. The sales curve corresponding to the most probable return level of 12 percent reaches these same levels of capacity after 4.5 and 7 years respectively. Another way of looking at the difference between the two sales curves is that after 4.5 years, when the most probable estimate of the rate of capacity utilization is 50 percent, the sales rate consistent with the 10 percent return curve is 35 percent of capacity. After 7 years, when the most probable estimate of sales is the full capacity of the plant, the sales rate consistent with the 10 percent return curve is about 70 percent of capacity.

These differences between the sales requirements—the area between the two curves—represent the "cushion against adversity" inherent in the project. It tells the decision maker that as long as actual sales do not fall below about 70 percent of the most probable level, the project can still meet the minimum profitability requirements of the company.

Of course, in this illustration there is a good chance that sales will fall below 70 percent of the most probable level. As Figure 9-1 indicates, there is a .3 probability that the rate of return of Project B will fall below 10 percent and a .15 probability that the return will fall below 8 percent. Figure 10-2 portrays this riskiness in terms of the actual marketing challenges that must be met if profitability aspirations are to be satisfied. Through such a portrayal, the decision maker can determine, for example, how many incremental millions of pounds of sales have to be "firmed up" by further market research if the risk is to be reduced to a tolerable level.

Another important dimension of project risk is illuminated by studying the effect of selling price on the required sales curve. Figure 10-3 shows the sales volumes required for a 10 percent rate of return under different selling prices (all other factors are identical to the base case).

The base case, with its decline in selling price from 30¢ to 28¢

Figure 10-3. Effect of selling price on sales volumes required for 10 percent DCF rate of return

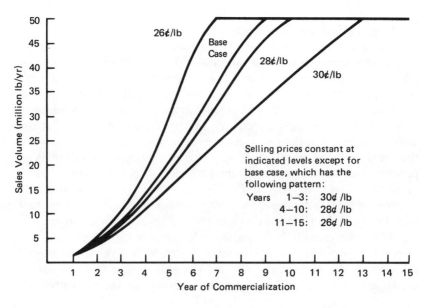

to 26¢ per pound, requires that plant capacity be reached in 9 years
to achieve a 10 percent DCF rate of return. A constant selling price
of 26¢ per pound requires that the plant be sold out in 7 years;
of 28¢ per pound, in 10 years; and of 30¢ per pound, in 13 years.
The sensitivity is similarly apparent in the wide spread between the
required sales curves in any given year. Thus, for a selling price
of 26¢ per pound, the plant must be sold out in year 7. For the
base case in that year, sales need only be at 70 percent of capacity;
for 28¢ per pound, at 65 percent of capacity; and for 30¢ per pound,
at only 50 percent of capacity.

This portrayal indicates the extent to which sales volume must
increase to offset a decline in selling price and thus achieve a 10
percent return. The relative differences between the curves would
not change appreciably with moderate changes in the return rate chosen.

In a similar vein, the sensitivity of sales volume to product lifetime
is shown in Figure 10-4. If the product lifetime is only 10 years
instead of 15 (all other factors being identical to the base case), capacity
operation must be reached after year 6 to achieve a 10 percent return.

Figure 10-4.   Effect of product lifetime on sales volumes required for 10
percent DCF rate of return

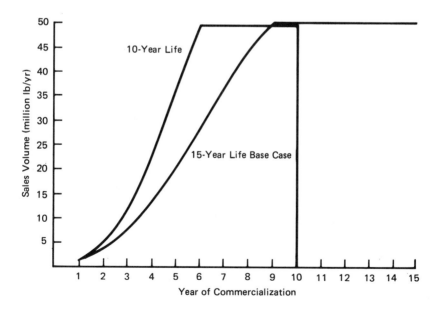

At this time, the base-case sales need reach less than 60 percent of plant capacity (having till year 9 to reach full capacity) in order to achieve the same return.

The examination of the effects of selling price and product lifetime on the sales volumes required to achieve a 10 percent rate of return is a form of sensitivity analysis (see Chapter 8) in which the impact on the rate of return of changes in the inputs are calculated. Here, however, the sensitivity is expressed in terms of the effect on the required sales pattern, the riskiest and most significant element in the analysis. Since the analyst is forced to consider explicitly the effects of alternative R&D strategies on the required marketing pattern, better decisions should result.

Figures 10-3 and 10-4, for example, emphasize and quantify the incentives for higher sales prices and extending product lifetime. They suggest that some additional R&D expenditures aimed at these goals could be justified. The technique permits a comparison of the costs and benefits associated with such a strategy.

To illustrate, the sensitivity of sales volume to manufacturing cost is examined in case 2, Figure 10-5. Case 2 differs from the base case in that it exhibits the following manufacturing cost–sales volume relationships:

$$MC = 14 + \frac{2}{S_t/C} \ \cent/lb \qquad \text{for} \qquad 0 \le S_t/C \le .25 \qquad (9a)$$

$$MC = 10 + \frac{3}{S_t/C} \ \cent/lb \qquad \text{for} \qquad .25 \le S_t/C \le 1 \qquad (9b)$$

The difference between the manufacturing costs specified by Equations (9a) and (9b) and those specified by base-case Equations (6a) and (6b) is a constant 2¢ per pound over the entire capacity range.

There is considerable incentive to somehow reduce manufacturing costs by 2¢ per pound. This incentive resides in the lower degree of risk inherent in case 2, as expressed by the lower sales volumes required to justify the project at the 10 percent level. Thus case 2 allows about half again as much time as the base case for sales to reach plant capacity—13 years versus 9 years—if equivalent rates of return are to be achieved. The "marketing latitude" inherent in case 2 suggests that it might be worthwhile to develop a lower-cost, more efficient process, even at the expense of additional R&D and/or capital investment spending.

Experimental evidence may indicate, for example, that a process

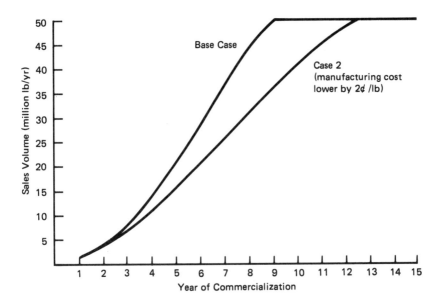

Figure 10-5. Effect of manufacturing cost on sales volumes required for 10 percent DCF rate of return

could be developed utilizing a less expensive raw material or providing a higher yield of on-specification product. If it is estimated that development of such a process would require spending an additional $800,000, or one-fourth more, on R&D, the projected R&D spending pattern in this instance—case 3—would be:

Year	R&D Spending
1968	$1,000,000
1969	1,000,000
1970	700,000
1971	500,000
1972	400,000
1973	400,000

Alternatively, there may be no anticipated increase in R&D spending associated with the development of a lower-cost process, but the required capital investiment may be higher by $1 million, or one-third—case 4. Finally, the development and implementation of the lower-cost process may require spending both one-fourth more in R&D and one-third more in capital investment—case 5. How much incentive would there be to undertake the necessary added expenditures for each of these possibilities?

The incentives are expressed in Figure 10-6 in terms of the sales volumes required to achieve a 10 percent DCF rate of return. The sales curves indicate considerable incentive to undertake process optimization at the costs indicated. Even case 5 has about two years' latitude in reaching plant capacity relative to the base case; when the base case must achieve plant capacity, case 5 need only have attained 85 percent of capacity. The decision maker can thus balance the reduction in marketing risk inherent in the lower sales requirements against the added costs that must be incurred early in the project life to achieve this reduction, and decide which is the preferred route.

The usefulness of MSA can be illustrated by considering other typical decisions that arise during new product development. For instance, the firm may want to offer more than one grade of a product for different industries. The second grade may have a higher purity, different stabilization, or different physical form. It is estimated that development of the new grade will require an additional $1 million

Figure 10-6. Effect of process optimization on sales volumes required for 10 percent DCF rate of return

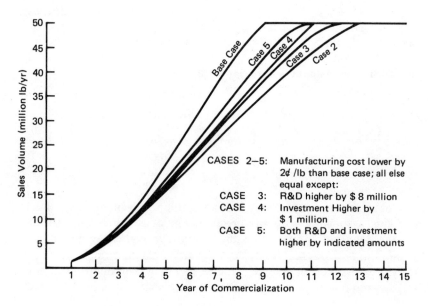

in R&D (all other factors remain identical to the base case). The projected R&D spending pattern in this instance—case 6—would be:

Year	R&D Spending
1968	$1,000,000
1969	900,000
1970	700,000
1971	600,000
1972	600,000
1973	400,000

The results of this increase in R&D spending are depicted in Figure 10-7. Case 6 must reach full capacity a year earlier than the base case—and must attain annual sales volumes some 15 percent higher than the base case until then—in order to achieve the same required rate of return. The critical question is: Will the presence of the second grade of product generate the increased sales volumes required to achieve the minimum acceptable return?

An analogous situation can be posed with regard to investment.

Figure 10-7. Effect of increased R & D costs on sales volumes required for 10 percent DCF rate of return

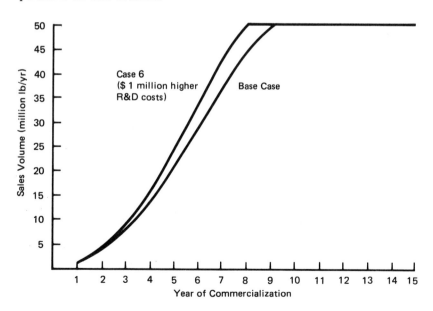

Assume that the new grade is simply a different physical form of the product. For example, the product could be offered as a fine powder, as coarse granules, or as a solution in various diluents. No additional R&D expenses are necessary, but the required investment is increased by $1 million—case 7—due to the need for additional equipment such as blending tanks and storage facilities. The cases are compared in Figure 10-8. Once again, the increased sales required by the added expenditure can be weighed against the potential contribution to sales offered by the new grade.

It is obvious that no amount of market research will permit the firm, from the vantage point of year 1 (or year − 3, say), to forecast sales so closely as to be able to distinguish between the sales curves required for the base case and case 6 or 7, for instance. However, the *absolute* level of sales is not the important consideration here; at issue is the *incremental* sales requirement associated with the tactic under consideration. This increment can be rationally considered even when the absolute sales level is in question.

Figure 10-8.   Effect of increased investment on sales volumes required for 10 percent DCF rate of return

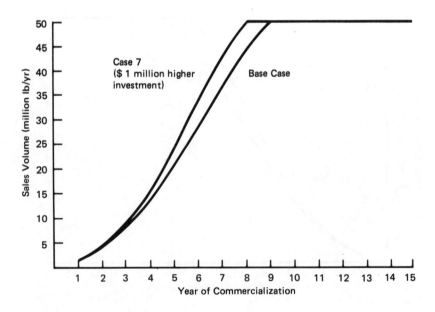

MSA, however, offers no automatic rules. Going back to Figure 10-6, for example, we can compare the required sales volumes associated with the base case and with case 5. From the marketing standpoint, the reduction in required sales shows that it would be less risky to spend an additional $.8 million on R&D and an additional $1 million on investment, to effect a 2¢-per-pound saving in manufacturing costs. A measure of the degree of risk thereby avoided is given by the area between the sales curves for the base case and for case 5.

This reduction in marketing risk must be balanced, against the risk involved in laying out an additional $1.8 million in the very early years of the project. If the added expenditure is undertaken and the product is an early failure, there will be an additional loss of $1.8 million. On the other hand, if the product is a success and the added expenditure is not made, the return will be considerably less, and the firm will be in a less favorable position to withstand competitive actions.

MSA provides only a quantitative framework for considering the risks and costs associated with alternative strategies. The decision, as always, depends on the judgment and risk preference of the decision maker.

chapter 11

Implications for R&D Management

Managements of companies in different industries differ markedly
in their opinion of how R&D can profitably be used. They do not
believe, for example, that what is right for the chemicals and electronics
industries is necessarily right for the petroleum-refining or steel
industries. Therefore, the role of R&D in a company's planning (and
in its future) varies widely from industry to industry and, within
the same industry, from company to company.

Corporate management has four basic strategic options to consider
and reconsider:

Harvest the present business.
Grow the business in its present form.
Create a new business or businesses (diversification).
Disinvest.

In the implementation of each of these options, the role of R&D and the level of R&D expenditures will be quite different.

Strategic planning for R&D must begin with an in-depth analysis of the company's performance (particularly as measured by the available cash flow) and the company's ability to achieve its business goals. R&D budgets must then be related to the specific stage the company has attained in its evolutionary development. Obviously, in application this principle is neither straightforward nor logical. As a result, R&D budgets typically emerge from an adversary relationship between corporate management and the management responsible for R&D.

Not surprisingly, many firms make significant changes in R&D funding and organization only when forced to by events outside their control, such as the rapid inflationary cycle beginning in 1974. In these periods corporate management always makes it clear that the return from R&D must be higher than the return the company could earn by investing in other productive facilities. An R&D manager's options in meeting this goal are limited by the present and possible financial performance of the company. The manager knows that R&D funding is a discretionary, changeable expenditure. The key is knowing the acceptable alternatives of when to change and how.

The critical strategic characteristic of R&D is the historic fact that only a small percentage of research starts can realistically be expected to lead to commercial success. Hence the importance of both the R&D manager's selection process and the need to insure that R&D objectives are consistent with business goals and purposes. The manager's portfolio of R&D activities must be sufficiently large and varied to allow for the expected high attrition rate.

Termination of R&D activities on rational grounds must be accepted as a fact of research life. Those activities that survive the early, low-expense stages must be critically assessed before larger expenditures are committed to them. Alternatives need to be carefully compared before a final selection is made.

These conclusions are not particularly remarkable or new. To remove them from the category of generality, however, we need to invest them with quantitative substance. What constitutes, for example, "rational grounds," "critical assessment," "careful comparisons"? Some concrete decision-making criteria are clearly required for the rational allocation of limited resources to R&D activities.

Unfortunately, the application of quantitative financial techniques to R&D decision making and control has not had particularly striking results. Many R&D managers have approached this subject with great caution. Others, at the opposite pole, have perpetrated abuses in applying such techniques and have discovered that the control of R&D costs through regular business accounting indicators—profits to sales, return on investment, cash flow, present value, and so forth—is unsatisfactory and confusing. As a result, only in R&D activities does so much money still ride each year on decisions based largely on judgment and faith rather than on quantitative dollars-and-cents analysis.

Knowing how to appraise investment opportunities is critical for successful interaction with corporate management. The fact that a potential business opportunity is founded on R&D has no influence on the mechanics of the analysis. Evaluation of the overall impact of the opportunity on the company—that is, a business evaluation—is required. The key to a successful R&D project is an analysis that indicates beforehand what to expect from the project and what the acceptable alternatives are. In addition, the analysis should help insure that R&D management and corporate management do indeed understand what they are doing and why.

Analysis has its pitfalls and can sow the seeds of future disasters. The pitfalls can be avoided by:

Understanding the complexities involved in critical decisions rather than just becoming a computer of elegant one-point answers.

Seeking out both strengths and weaknesses in an R&D proposal.

Distinguishing properly between accounting information and business economics—that is, critically evaluating all sorts of financial data.

Putting all estimates on an enterprising basis and not simply attempting to accumulate safety factors everywhere.

Establishing a project opportunity team and not just acting as a go/no-go evaluator.

Successful analysis depends on an increased flow of information from the marketplace. Only with the use of market information can the risk inherent in R&D decision making be reduced.

R&D's link with the marketplace is, of course, the company's marketing organization. Just what is involved in strengthening the R&D–marketing bond? It is necessary that marketing be closely integrated with R&D from an early point in the process—certainly before development work begins. Such integration can be achieved physically, through organization structure and procedural mechanisms.

It can also be achieved philosophically. Indeed, a commitment to philosophy must accompany (or precede) any organizational or procedural implementation.

One of the foremost requirements is a commitment by R&D to the "fact of corporate life": the *real* object of research and development is sales. This statement may seem too obvious to dwell on. But, for all the lip service paid to it, it is not universally accepted.

Evidence for a possible lack of interest in sales can be seen in such a seemingly innocuous area as terminology. In many firms R&D activities are categorized (following the National Science Foundation classifications) as basic research, applied research, and development. In recognition of the real objective of R&D, research activities should be more accurately categorized by their business *purpose*. Although managers then cannot identify the "basic research" costs or the "applied research" costs, they know the costs and their distribution among various business purposes.

Terminology is not simply a matter of convention. It indicates the company's attitude toward research, signifying a commitment, or lack thereof, of R&D objectives to business purpose. Such a commitment has a great significance for R&D productivity, which is, after all, meaningful (and measurable) only in terms of the attainment of business objectives.

The integration of R&D and marketing must extend beyond philosophical commitments and terminology, of course. It can play a major role in the organization of the R&D effort. In the procedural area, the integration of R&D and marketing implies conducting market research studies alongside the development effort. For a new product development project, for example, an early assessment must be made of the markets in which the product will be sold, along with the probable penetration and selling price. Such an analysis must take place as soon as the product concept has been identified and before the costly technical development phase begins. Assessment must continue throughout the development process to make sure that the evolving characteristics of the product will provide the required competitive edge and market appeal. Competitive developments must therefore be monitored and taken into consideration.

The procedure summarized so neatly is not so easily done. To begin with, customers must be met as soon as possible. This introduces expenditures and all sorts of risks. Clearly, there are no simple answers. But one thing is certain: the costs of product failures are greater today than ever before, and the record indicates that the reason for most new product failures is inadequate understanding of market forces. Unfortunately, despite the crucial importance of market knowledge and other nontechnical considerations to R&D decision making, it is marketing information that is the most difficult to obtain. It is tempting to use the difficulty of obtaining information as an excuse for neglecting the marketing area. Indeed, many R&D managers may do this, since they are most likely technically trained people who are interested more in the tangible challenges of technical development than in the elusive problems of the marketplace. Managers possessed of a new piece of research knowledge can find it extraordinarily difficult to establish and validate its relevance to any economic need.

Business objectives—which call for elaborate market knowledge and other nonscientific considerations—must be used to provide the final target and the intermediate directional markers along the development path. Managers must always remember that the profitability of their R&D effort is critically dependent on many actions and decisions in completely nonscientific areas. A new scientific discovery is really only an admission ticket to the big game.

Companies must continually reevaluate the financial justification for their development projects. Analysis must be undertaken before the development effort is initiated and must be brought up to date regularly as the various elements come into sharper focus. This is "success analysis." Its aim is to answer the question "If R&D is successful, then what?" Such an analysis is much more than a conventional marketing investigation; it involves examining all the business elements that will effect the conversion of R&D activities into revenue and profits. On occasion the analysis should significantly change the direction of the R&D project and contribute to its profitable outcome.

A rational approach to the financial analysis of development projects involves building a business model for the new product (or process) along the same accounting lines used in analyzing all other company operations (such as requests for capital investment appropriations). Into the model go estimates for plant costs, working capital requirements, manufacturing costs, selling expenses, R&D costs, selling prices, sales volumes, and so forth. The timing of the various inflows and outflows also must be estimated. To allow for risk, the values of the elements can be described in terms of their subjective probability distributions. These values are then used to calculate the probability distributions of DCF rates of return (or of present worths).

The most probable return can then be compared with a target return level to determine whether the project offers (in light of its risk, as measured by the variance of the distribution of possible returns) sufficient incentive to continue development. This is a lot of work, even without the probability distributions.

However, managers must guard against any mechanical formula (even a DCF criterion) for "evaluating" R&D projects. All the formulas still in vogue invoke some arbitrary rating system or other standard that reduces their objectivity and makes them incompatible with the criteria used to evaluate capital investments. The presence of risk heightens, rather than detracts from, the need to conduct venture analyses of development projects. The analysis forces those responsible to make explicit assumptions about markets and marketing, timetables, competition, and economic goals. The results permit R&D management to make rational judgments about complex, multivariable problems. They also focus attention on the information needed to keep the technical and marketing risks in step. A profitability estimate in the form of a venture analysis at an early stage of development may be little more than a statement of hope, but it also provides guidance as to how that hope may best be realized.

notes

This book has drawn upon a great deal of published literature. A complete listing of references or a bibliography of the relevant literature would run to several hundred if not a thousand items. The scope of the literature is made clear by some of the bibliographies published during the past ten years in the *IEEE Transactions on Engineering Management* and by the critical reviews available from the Research Program in Industrial Economics at Case Western Reserve University.

Our objective in these Notes is to provide a roadmap to the published literature that will lead readers to all the references mentioned or implied in the text.

CHAPTER 1: THE NATURE OF RESEARCH AND DEVELOPMENT   Research and development expenditures and patterns of funding are discussed in many national and trade magazines. Unfortunately, this information is often neither comprehensive nor clear. The publications of the

National Science Foundation, particularly the latest issue of *National Patterns of R&D Resources: Funds and Manpower in the United States,* are the key.

In 1976 and 1977 *Business Week* published its survey of corporate research and development. (Hopefully, the pattern will continue in future years.) The survey supplements the NSF data by pinpointing the expenditures of major industrial concerns in the United States. *Business Week's* information is obtained from 10-K data and other corporate reports.

The publications and meetings of the Industrial Research Institute (IRI) are an excellent source of information about research funding and manpower. The institute's more than 200 corporate members meet twice a year to discuss ways to improve the management of R&D. They also hold workshops, seminars, and study groups on selected topics and publish a journal, *Research Management.* The offices of IRI are at 100 Park Avenue, New York, N.Y. 10017.

Soloman Fabricant and his associates at New York University carried out an extensive NSF study called "Accounting by Business Firms for Investment in R&D." A summary of this study, prepared by Shahid L. Ansari, is in IRI's *Research Management* (November 1976), pp. 28–33.

A discussion of the complex and currently unresolved conceptual issues that arise in research on innovation is given by George Downs, Jr. and Lawrence Mohr in *Administrative Science Quarterly* (December 1976), pp. 700–774. This review is also an excellent introduction to the innovation literature.

CHAPTER 2: THE RISKS AND BENEFITS OF RESEARCH AND DEVELOPMENT   A number of seminal works on the risks and benefits of R&D are excellent guides to the published literature. The three that we have found to be most helpful over the years are:

> Bela Gold, *Research Technological Change in Economic Analysis* (Lexington, Mass.: Lexington Books, 1977).

> Edwin Mansfield, *Industrial Research and Technological Innovation and Econometric Analysis* (New York: W. W. Norton, 1968).

> National Science Foundation, *Research and Development and Economic Growth Productivity,* NSF 72-303 (Washington, D.C., 1972).

The Bela Gold volume is particularly up to date and comprehensive in its discussion of the relevant literature. In various talks since the

NSF study was issued, Leonard Lederman has updated his analysis of the relationships between science, technology, and economic growth. Several of these updates have been published.

Over the years Arthur D. Little, Inc. has conducted numerous studies on the relationship between R&D expenditures and the financial performance of firms and industries. These studies, which are periodically updated, are widely circulated throughout industry. Unfortunately, the studies may not be readily retrievable.

The literature abounds with studies, surveys, and anecdotes about new product successes and failures. Many of the most relevant studies have been published by The Conference Board. Of particular value is a study by David Hopkins and Earl Bailey in *The Conference Board Record* (June 1971), pp. 12–24.

A complete listing of the studies used in developing Table 2-1 is given in C. Merle Crawford, *Journal of Marketing* (April 1977), pp. 52–61.

CHAPTER 3: STRATEGIC PLANNING AND RESEARCH AND DEVELOPMENT   There is a considerable body of literature that can be called planning literature, even though the field of planning is less clearly positioned than more mature disciplines. In addition, there is a vast amount of literature related to planning from such areas as policy theory, organization theory, systems theory and cybernetics, management control theory, management information systems theory, and computer modeling. The best introduction to this literature is Russell Ackoff, *A Concept of Corporate Planning* (New York: John Wiley and Sons, 1970). Peter Lorang of the Sloan School of Management, Massachusetts Institute of Technology, presented at the sixth annual meeting of the American Institute for Decision Sciences (1974) an excellent paper called "Formal Planning System: The State of the Art." However, the paper may not be readily available.

The product life cycle and portfolio concepts have been championed by Arthur D. Little, Inc. and the Boston Consulting Group, respectively. Relevant publications from these concerns are widely scattered and available only from them. However, an excellent review of these techniques appeared in the Spring 1975 issue of the *Journal of Contemporary Business* ("New Venture Planning and Market Entry"). The journal is published by the Graduate School of Business Administration, University of Washington, Seattle, Wash. 98195.

The concepts of cash flow, funds flow, and income flow for a corporation and their relationship to the planning structure are discussed in depth in Robert Jaedicke and Robert Sprouse, *Accounting Flows, Income Funds and Cash* (Englewood Cliffs, N.J.: Prentice-Hall, 1965). The Modigliani-Miller hypothesis was first discussed in their article "Cost of Capital, Corporation Finance, and Theory of Investment," *American Economic Review* (June 1958), p. 45. This article has been reprinted in Ezra Solomon, ed., *The Management of Corporate Capital* (New York: Free Press, 1959). The traditional position on the choice between debt and equity is discussed fully in Ezra Solomon, *The Theory of Financial Management* (New York: Columbia University Press, 1963), pp. 92 ff.

The equations relating the price/earnings ratio to corporate return, dividend payout, and growth for a stockholder's return of 10 percent are given below:

$$\frac{\text{Price}}{\text{Earnings}} = \frac{100\,(\%\text{ corporate return} - \%\text{ growth in dividends})}{\%\text{ corporate return }(10 - \%\text{ growth in dividends})}$$

$$\frac{\text{Price}}{\text{Earnings}} = \frac{100\,(\%\text{ dividend payout})}{1{,}000 - \%\text{ corporate return }(100 - \%\text{ dividend payout})}$$

The corporate return used in these equations is net return on net worth.

Anyone who wishes the specific derivation of this equation can contact the authors. The derivation is similar to those in Alexander Robichel and Stewart Meyer, *Optimal Financing Decisions* (Englewood Cliffs, N.J.: Prentice-Hall, 1965).

CHAPTER 4: THE BASIC ECONOMIC STRUCTURE OF A BUSINESS   A number of books discuss the tools and techniques of financial analysis. Some of them are designed for self-teaching. One of the most valuable in our judgment—because of its scope and clarity—is J. Fred Weston and Maurice B. Goudzwaard, eds., *The Treasurer's Handbook* (Homewood, Ill.: Dow Jones–Irwin, 1976). Also valuable is Prentice-Hall's Foundation of Finance Series, edited by Ezra Solomon. This series covers the major components of financial management in short, independent books. Each volume is written by someone outstanding in the field and focuses on a single subject area. Before you do anything else, get the latest copy of the free booklett *How to Read a Financial Report*, from Merrill Lynch, Pierce, Fenner & Smith, Inc. This booklet explains in depth the concepts utilized in the first half of the chapter.

The cost of capital concept is reviewed in Robert Anthony,

Accounting for the Cost of Interest (Lexington, Mass: Lexington Books, 1975). This book is a polemic in that it proposes to build cost of capital into all business decisions; it also presents an outstanding review of the literature.

Lawrence Fisher and James Lorie, *A Half-Century of Returns on Stocks and Bonds* (Chicago: University of Chicago Press, 1977), present extensive data on rates of return decision for the period 1926–1976. Periodically, Oppenheimer and Co., New York, N.Y., issues both historical studies and projections of expected returns from stocks, bonds, and Treasury bills.

CHAPTER 5: THE APPRAISAL OF ALTERNATIVES   All the concepts discussed in this chapter are presented in depth (by people who have used the techniques day in and day out) in J. Fred Weston and Maurice B. Goudzwaard, eds., *The Treasurer's Handbook* (Homewood, Ill.: Dow Jones–Irwin, 1976). To learn some of the tricks we did not talk about, look at Harold Biermann, *Quantitative Analysis for Business Decisions,* 3rd ed. (Homewood, Ill.: Richard D. Irwin, 1969).

CHAPTER 6: THE DISCOUNTED CASH FLOW METHOD OF ECONOMIC ANALYSIS   The one essential book on DCF analysis is Harold Biermann and Seymour Smidt, *The Capital Budgeting Decision* (New York: Macmillan, 1971). Chapter 17 of J. Fred Weston and Maurice B. Goudzwaard, eds., *The Treasurer's Handbook* (Homewood, Ill.: Dow Jones–Irwin, 1976), gives extensive references, some of which can be extremely helpful.

If you are going to do DCF calculations on a regular basis, you are going to need some form of computer. For many purposes the desk-top Hewlett Packard HP-92 is ideal. It can solve problems involving time and money, compound-interest balloons, internal rates of return for 30 uneven cash flows, and so on; it also provides a printout of all calculations. Our experience with it has been outstanding.

If you move on to venture analysis, you will need some computer software. The best available in our judgment is SIMPLAN, a product developed by Social Systems Inc. and offered as a service by Informatics Inc., Fairfield, N.J. SIMPLAN was developed specifically for the formulation and programming of corporate planning models. Its report generator is outstanding, and it can be used easily by individuals who have no previous experience in computer modeling.

The impact of inflation on the rate of return from investment has generated considerable interest in the literature. Most of the articles are difficult to follow. In our judgment, the best article on the subject is that by Brant Allen in *Business Horizons* (December 1976), pp. 30–39. Allen also provides an excellent reference list.

CHAPTER 7: THE ELEMENTS OF VENTURE ANALYSIS The time elements involved in moving from R&D to commercialization are covered by Albert Brown in *ChemTech* (December 1973), pp. 709–713. Bela Gold, *Research Technological Change in Economic Analysis* (Lexington, Mass: Lexington Books, 1977), also contains valuable information on this subject.

Marketing and the marketing forecast must be made specific to the activity, company, industry, and products being studied. There is no simple way to do this. However, we feel that Bill Butler and Bob Tavishes, *How Business Economists Forecast* (Englewood Cliffs, N.J.: Prentice-Hall, 1966), should be on everyone's shelf. Then add those of the following that seem relevant:

Robert Ayres, *Technological Forecasting and Long-Range Planning* (New York: McGraw-Hill, 1969).

Robert Brown, *Smoothing, Forecasting, and Prediction of Discrete Time Series* (Englewood Cliffs, N.J.: Prentice-Hall, 1963).

John Carlson, "Forecasting Errors and Business Cycles," *American Economic Review* (June 1967), pp. 462–481.

Roger Chisholm and Gilbert Whitaker, Jr., *Forecasting Methods* (Homewood, Ill.: Richard D. Irwin, 1971).

Lawrence Fisher, "The Methodology of Long-Term Forecasting of Demand for Industrial Products," *Industrial Marketing Research Association Journal* (February 1969), pp. 3–47.

Ivan Friend and Peter Taubman, "A Short-Term Forecasting Model," *Review of Economics and Statistics* (August 1964), pp. 229–236.

F. Thomas Juster, "Consumer Buying Intentions and Purchase Probability: An Experiment in Survey Design," *Journal of the American Statistical Association* (September 1966), pp. 658–696.

Robert Kirby, "Comparisons of Short- and Medium-Term Forecasting Techniques," *Management Science* (December 1966), pp. B202–B210.

Lawrence Klein, *An Introduction to Econometrics* (Englewood Cliffs, N.J.: Prentice-Hall, 1962).

Jacob Mincer, *Economic Forecasts and Expectations: Analysis of Forecasting Behavior and Performance* (New York: Columbia University Press, 1969).

National Industrial Conference Board, *Forecasting Sales*, Studies in Business Policy No. 106 (1963).

Investment and operation cost estimates are given only generalized treatment in the literature. Moreover, specific operating cost and investment information is usually out of date by the time it appears in the literature. Various McGraw-Hill publications, particularly *Chemical Engineering* magazine, present valuable guidelines on making these estimates and suggestions on the information that must be collected to make adequate estimates. Of particular value is K. M. Guthrey's article in *Chemical Engineering* (March 24, 1969), pp. 114–142, which introduces a modular technique for making fast, accurate, and consistent cost estimates.

The learning curve is an important concept for R&D managers. An excellent reference on the subject is *Perspectives on an Experience* (1970), published by the organization that popularized the concept, the Boston Consulting Group. David Brook's article in *Technology Review* (March–April 1976), pp. 53–59, is also helpful and provides some valuable suggested readings.

All the other elements needed to make up a venture analysis are covered in J. Fred Weston and Maurice B. Goudzwaard, eds., *The Treasurer's Handbook* (Homewood, Ill.: Dow Jones–Irwin, 1976).

CHAPTER 8: VENTURE ANALYSIS   If you are going to do venture analysis, you must have a copy of *Du Pont Guide to Venture Analysis: A Framework for Venture Planning*. It is available from E. I. Du Pont de Nemours & Company, Inc., Education and Applied Technology Division, Wilmington, Del. 19898. The book provides a detailed, step-by-step guide to venture analysis, as well as valuable references and ideas about risk analysis (Chapter 9).

You should also become acquainted with the publications of Jack Malloy of Amoco. Of particular interest is his article in *Chemical and Industry* (October 30, 1971), pp. 1242–1249, which discusses how venture analysis can be used in various situations; and his article on projecting chemicals prices in *Chemical Engineering Progress* (September 1974), pp. 77–80.

James Weaver's article on the ICI United States system for monitoring capital expenditure in the *Engineering Economist* (Fall

1974), pp. 1–35, is particularly valuable in showing how the concepts discussed in the chapter can be applied in an ongoing management planning and control program.

CHAPTER 9: RISK AND UNCERTAINTY   The best introduction to risk and uncertainty are two Penguin paperbacks: Gordon M. Kaufmann and Howard Thomas, *Modern Decision Analysis* (1977), and T. J. Moore and Howard Thomas, *Anatomy of Decisions* (1977). Some of the more readable references on decision analysis and simulation are summarized below:

R. V. Brown, A. S. Kahr, and C. Peterson, *Decision Analysis for the Manager* (New York: Holt, Rinehart and Winston, 1974).

R. E. Frank and P. E. Green, *Quantitative Methods in Marketing* (Englewood Cliffs, N.J.: Prentice-Hall, 1967).

C. J. Grayson, Jr., "Decisions Under Uncertainty: Drilling Decisions by Oil and Gas Operators" (Cambridge, Mass.: Harvard School of Business, 1960).

Harvard Business Review, Statistical Decision Series, Parts I–IV (Boston, Mass., 1951–1970).

J. F. Magee, "How to Use Decision Trees in Capital Investment," *Harvard Business Review* (September–October 1964), pp. 79–96.

The optimum plant size problem is discussed extensively with particular application to the electric utilities industry by Sam Peck in the *Bell Journal of Economics and Management Science* (Autumn 1974), pp. 420–460. L. M. Rose has made a preliminary extension of the manufacturing capacity problem to an uncertain demand in *Engineering and Process Economics* (September 1977), pp. 17–25. His references are particularly helpful, and he makes it clear that the problem is a complex one.

Computer software is mandatory for risk analysis. We have found that from Social Systems, Inc., Chapel Hill, N.C., to be quite satisfactory.

CHAPTER 10: THE MARKET SPECIFICATION APPROACH   The computer program referred to in this chapter is available on request from the authors.

CHAPTER 11: IMPLICATIONS FOR R&D MANAGEMENT   We recommend Lowell Steele, *Innovation in Big Business* (New York: Elsevier, 1975).

index